Introduction to Statistics: Analyzing Data

Introduction to Statistics: Analyzing Data

Nash Stokes

MURPHY & MOORE
www.murphy-moorepublishing.com

Introduction to Statistics: Analyzing Data
Nash Stokes
ISBN: 978-1-63987-336-4 (Hardback)

MURPHY & MOORE
Published by Murphy & Moore Publishing,
1 Rockefeller Plaza,
New York City, NY 10020, USA

Cataloging-in-Publication Data

Introduction to statistics : analyzing data / Nash Stokes.
p. cm.
Includes bibliographical references and index.
ISBN 978-1-63987-336-4
1. Statistics. 2. Mathematical statistics. 3. Statistics--Methodology.
I. Stokes, Nash.
QA276 .P73 2022
519.5--dc23

For more information regarding Murphy & Moore Publishing and its products, please visit the publisher's website www.murphy-moorepublishing.com

Table of Contents

Preface

The sub-discipline of mathematics which is concerned with the collection, interpretation and analysis of data is known as statistics. It is further classified into two sub-branches, namely, descriptive statistics and inferential statistics. Descriptive statistics is a form of non-parametric statistics. It makes use of various tools such as measures of central tendency including mean, median, mode, and measures of dispersion. Kurtosis and skewness are also studied within the domain of this subject. Inferential statistics applies the principles underlying distribution of probability. Different types of assumption models studied under this discipline are classified into fully parametric, non-parametric and semi-parametric. Statistics finds extensive application in the fields of computer science, economics, biology, machine learning and data mining. This book provides significant information of this discipline to help develop a good understanding of statistics and related fields. It is compiled in such a manner, that it will provide in-depth knowledge about the theory and practice of this field. Coherent flow of topics, student-friendly language and extensive use of examples make this book an invaluable source of knowledge.

A short introduction to every chapter is written below to provide an overview of the content of the book:

Chapter 1 - Statistics is the discipline that is concerned with gathering, reviewing and interpreting numerical data in large quantities. The topics like descriptive statistics, inferential statistics, variables, measures of center and measures of variation are discussed in this chapter. This is an introductory chapter which introduces the reader to the field of statistics; **Chapter 2** - Probability is the branch of mathematics which deals with numerical descriptions of the chances of an event to occur. This chapter explains functions, sequences, limit and continuity, random variables, world of integrals, distribution functions, etc. This chapter has been carefully written to provide an easy understanding of the varied facets of probability; **Chapter 3** - The process of using data analysis to conclude the parameters of populations or scientific truths on the basis of random sampling is known as statistical inference. This chapter elaborates nonparametric inference, bayesian inference, point estimation, criteria of estimation, unbiased estimation, etc. All the diverse principles of statistical inference and estimation have been carefully analyzed in this chapter; **Chapter 4** - The process of selecting a statistical sample which is a subset of individuals from a statistical population to determine the characteristics of the whole population is known as sampling. This chapter encompasses topics such as sampling, census, sampling frame, survey methodology and statistical unit. It closely determines the key concepts of sampling to provide an extensive understanding of the subject; **Chapter 5** - The branch of statistics which deals with accumulation and interpretation of data to uncover the patterns and trends is known as statistical analysis. This chapter unfolds types of data, primary data, secondary data, quantitative and qualitative data, census method and survey method. These topics are crucial for a complete understanding of the subject; **Chapter 6** - The set of statistical methods used to estimate the relationship between dependent and independent variables is known as regression analysis. This chapter discusses in detail topics such as linear regression analysis, nonlinear regression, nonparametric regression, semiparametric regression and regression models; **Chapter 7**- The collection of random variables that are guided by a mathematical set which means every random variable is incomparably associated with an element in the set is known as stochastic process. This chapter includes Wiener process, Ornstein-Uhlenbeck process, Random walk and Poisson point process. It will help the reader in gaining a better perspective about the stochastic process.

I extend my sincere thanks to the publisher for considering me worthy of this task. Finally, I thank my family for being a source of support and help.

Nash Stokes

1

Basics of Statistics

Statistics is the discipline that is concerned with gathering, reviewing and interpreting numerical data in large quantities. The topics like descriptive statistics, inferential statistics, variables, measures of center and measures of variation are discussed in this chapter. This is an introductory chapter which introduces the reader to the field of statistics.

Statistics is the science of collecting, analyzing, presenting, and interpreting data. Governmental needs for census data as well as information about a variety of economic activities provided much of the early impetus for the field of statistics. Currently the need to turn the large amounts of data available in many applied fields into useful information has stimulated both theoretical and practical developments in statistics.

Data are the facts and figures that are collected, analyzed, and summarized for presentation and interpretation. Data may be classified as either quantitative or qualitative. Quantitative data measure either how much or how many of something, and qualitative data provide labels, or names, for categories of like items. For example, suppose that a particular study is interested in characteristics such as age, gender, marital status, and annual income for a sample of 100 individuals. These characteristics would be called the variables of the study, and data values for each of the variables would be associated with each individual. Thus, the data values of 28, male, single, and $30,000 would be recorded for a 28-year-old single male with an annual income of $30,000. With 100 individuals and 4 variables, the data set would have 100 × 4 = 400 items. In this example, age and annual income are quantitative variables; the corresponding data values indicate how many years and how much money for each individual. Gender and marital status are qualitative variables. The labels male and female provide the qualitative data for gender, and the labels single, married, divorced, and widowed indicate marital status.

Sample survey methods are used to collect data from observational studies, and experimental design methods are used to collect data from experimental studies. The area of descriptive statistics is concerned primarily with methods of presenting and interpreting data using graphs, tables, and numerical summaries. Whenever statisticians use data from a sample—i.e., a subset of the population—to make statements about a population, they are performing statistical inference. Estimation and hypothesis testing are procedures used to make statistical inferences. Fields such as health care, biology, chemistry, physics, education, engineering, business, and economics make extensive use of statistical inference.

Methods of probability were developed initially for the analysis of gambling games. Probability plays a key role in statistical inference; it is used to provide measures of the quality and precision of the inferences.

Descriptive Statistics

Descriptive statistics are brief descriptive coefficients that summarize a given data set, which can be either a representation of the entire or a sample of a population. Descriptive statistics are broken down into measures of central tendency and measures of variability (spread). Measures of central tendency include the mean, median and mode, while measures of variability include standard deviation, variance, minimum and maximum variables, and kurtosis and skewness.

Descriptive statistics, in short, help describe and understand the features of a specific data set by giving short summaries about the sample and measures of the data. The most recognized types of descriptive statistics are measures of center: the mean, median and mode, which are used at almost all levels of math and statistics. The mean, or the average, is calculated by adding all the figures within the data set and then dividing by the number of figures within the set. For example, the sum of the following data set is 20: (2, 3, 4, 5, 6). The mean is 4 (20/5). The mode of a data set is the value appearing most often, and the median is the figure situated in the middle of the data set. It is the figure separating the higher figures from the lower figures within a data set. However, there are less common types of descriptive statistics that are still very important.

People use descriptive statistics to repurpose hard-to-understand quantitative insights across a large data set into bite-sized descriptions. A student's grade point average (GPA), for example, provides a good understanding of descriptive statistics. The idea of a GPA is that it takes data points from a wide range of exams, classes and grades, and averages them together to provide a general understanding of a student's overall academic performance. A student's personal GPA reflects their mean academic performance.

Measures of Descriptive Statistics

All descriptive statistics are either measures of central tendency or measures of variability, also known as measures of dispersion. Measures of central tendency focus on the average or middle values of data sets, whereas measures of variability focus on the dispersion of data. These two measures use graphs, tables and general discussions to help people understand the meaning of the analyzed data.

Measures of central tendency describe the center position of a distribution for a data set. A person analyzes the frequency of each data point in the distribution and describes it using the mean, median or mode, which measures the most common patterns of the analyzed data set.

Measures of variability, or the measures of spread, aid in analyzing how spread out the distribution is for a set of data. For example, while the measures of central tendency may give a person the average of a data set, it does not describe how the data is distributed within the set. So while the average of the data may be 65 out of 100, there can still be data points at both 1 and 100. Measures of variability help communicate this by describing the shape and spread of the data set. Range, quartiles, absolute deviation and variance are all examples of measures of variability. Consider the following data set: 5, 19, 24, 62, 91, 100. The range of that data set is 95, which is calculated by subtracting the lowest number 5 in the data set from the highest 100.

Inferential Statistics

Inferential statistics is used to analyse the results and draw conclusions. Experts described inferential statistics as the mathematics and logic of how this generalization from sample to population can be made. These procedures might be used to estimate the likelihood that the collected data occurred by chance and to draw conclusions about a larger population from which samples were collected. Inferential statistics helps us answer the following questions:

- Making inferences about a population from a sample.
- Concluding whether a sample is significantly different from the population. Interested in sports rather than watching television.
- If adding or removing a feature from a model will help in improving it.
- If one model is significantly different from the other.
- Hypothesis Testing.

This shows us why inferential statistics is important and why it is worth investing time and effort in learning these concepts.

Normal Distribution and z — Statistic

The normal distribution is also known as the bell curve having the following properties:

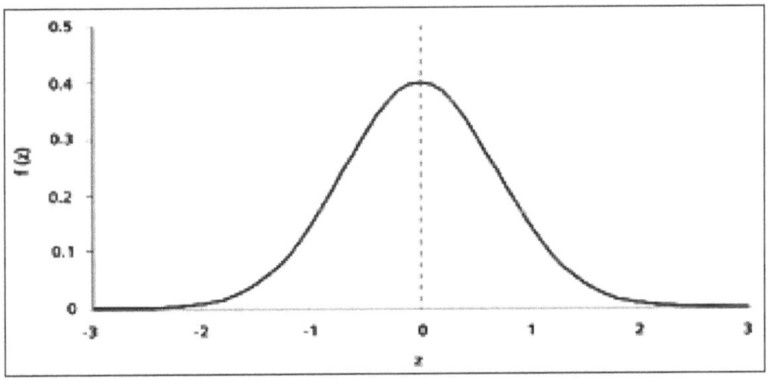

- Mean = median = mode.
- The curve is symmetric with half of the values on the left and half of the values on the right.
- The area under the curve is 1.

In a normal distribution:

- 68% of the data falls within one standard deviation of the mean.
- 95% of the data falls within two standard deviations of the mean.
- 99.7 % of the data falls within three standard deviations of the mean.

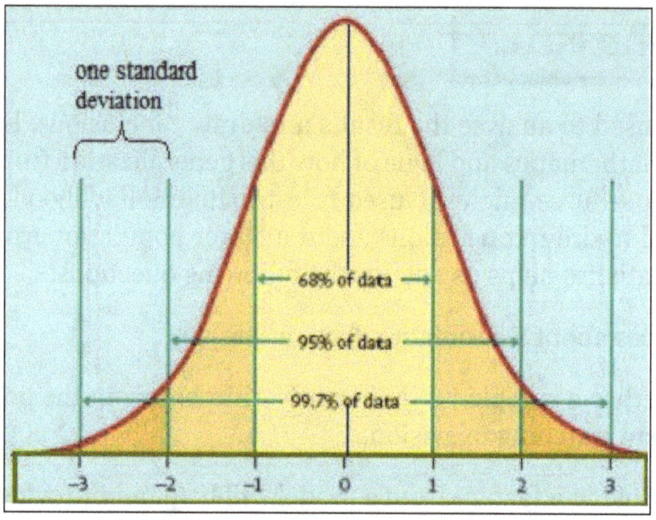

For calculating the probability of occurrence of an event we need the z — statistic. The formula for calculating the z — statistic is:

$$Z = \frac{x - \mu}{\sigma}$$

where x is the value for which we want to calculate the z — value. μ and σ are the population mean and standard deviation respectively. Basically what we are doing here is standardizing the normal curve by moving the mean to 0 and converting the standard deviation to 1. The z — statistic is essentially the distance of the value from the mean calculated in standard deviation terms. So a z — value of 1.67 means that the value is 1.67 standard deviations away from the mean in the positive direction. We then find the probability by looking up the corresponding z — value from the z table.

Central Limit Theorem

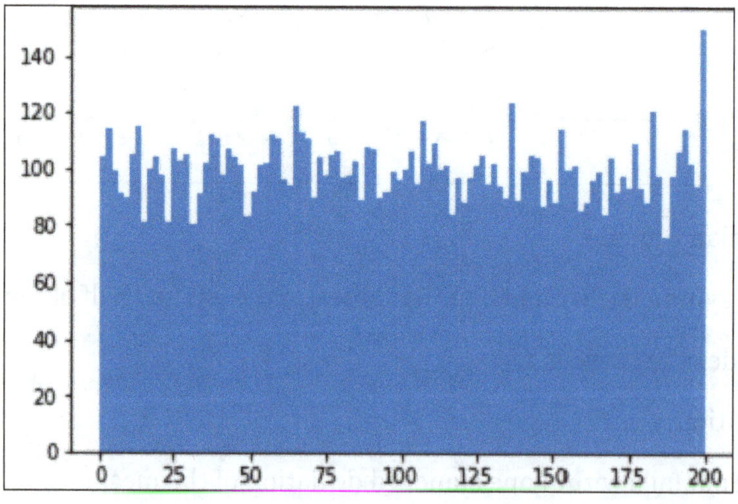

The main importance of the normal distribution comes from the central limit theorem. The central limit theorem states that the sampling distribution of the mean of sample means approaches the normal distribution as the sample size gets larger no matter the shape of the population distribution. Let's look at this through an example. Consider the case where we look at the number of tweets a person makes in a week (randomly generated data between 0 and 200). The frequency distribution of the data looks like this:

This is not similar to any kind of distribution we know. Now lets take 1000 random samples of size 50 from this data and calculate the mean of each sample. When we plot these means we get a normal distribution curve also known as the sampling curve or the sampling distribution.

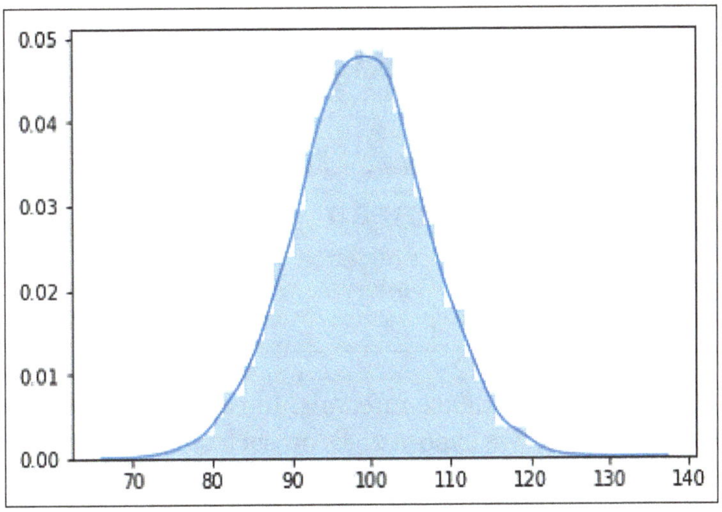

Mean = 98.78 (population mean = 98.87).

The central limit theorem has some important properties:

- The mean of the population is approximately equal to the mean of the sampling distribution. population mean (98.87) is approximately equal to the mean of sampling distribution (98.78).
- The standard deviation of the sampling distribution also known as the standard error is equal to the population standard deviation divided by the square root of the sample size. As a result the greater the sample size, the lower the standard deviation and greater accuracy in determining the sample mean from the population mean.

$$\sigma_{\text{sample means}} = \frac{\sigma_{\text{population}}}{\sqrt{N}}$$

- The distribution of sample means is normal regardless of the shape of the population distribution. It means that even if the original distribution is skewed or bimodal or some other distribution the mean of sample means is always a normal distribution. This is what makes the central limit theorem so powerful.

For the central limit theorem to hold the sample size should be sufficiently large (generally > 30).

Confidence Interval

How well the sample statistics estimates the underlying population is always an issue. A confidence interval addresses this issue as it provides a range of values which may likely contain the population parameter.

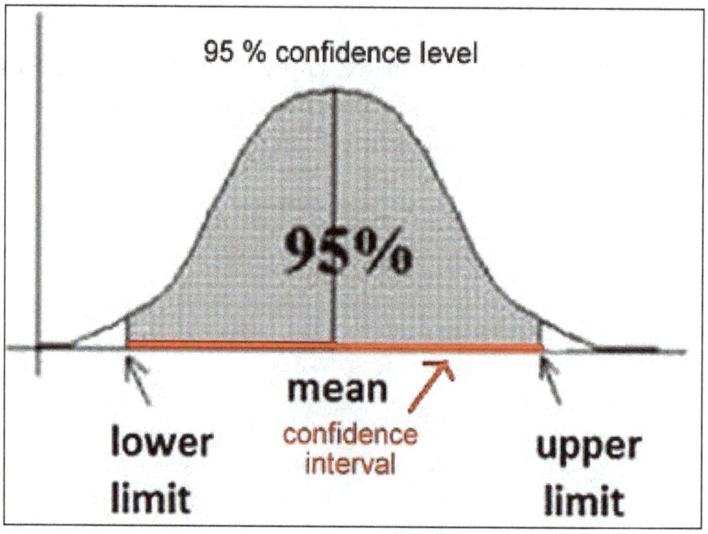

There are one sided or two sided confidence intervals. In two sided confidence interval, if we are choosing a 95% confidence interval, we choose 2.5% on each side of the tail and then calculate the range. In a one sided confidence interval we calculate the confidence interval by taking the entire 5% either to the left or to the right of the distribution. The above image displays a two sided confidence interval. The formula we use to calculate the confidence interval is:

$$C.I = \bar{X} \pm Z_{\alpha/2}\left(\sigma / (\sqrt{N})\right)$$

where the symbols stand for:

- $\bar{X}$: Sample Mean.
- $Z_{\alpha/2}$: Z value for the desired confidence level.
- α : Confidence level.
- σ : The standard deviation of the population.

Another important term in the confidence interval concept is the margin of error. It is half the size of a confidence interval. It is known as the sampling error and it means that if the sample mean is within the margin of error then its actual value is the population mean and the difference only occurs due to chance. Otherwise the results are considered to be statistically significant.

$$\text{margin of error} = Z * \left(\frac{\sigma}{\sqrt{N}}\right)$$

When we take a 95% confidence interval it does not mean that the population means is within the range we derive with a 95% chance. The confidence interval represents the frequency (i.e. proportion) of possible confidence intervals that contain the true value of the unknown population parameter. So if we take infinitely many samples and find the confidence interval range for each of these samples then the number of intervals that contain the population parameter is equal to the confidence interval. If we take a 95% confidence interval then it means that the population parameter is present in 95% of all possible confidence interval ranges.

Variables

In statistics, a variable has two defining characteristics:

- A variable is an attribute that describes a person, place, thing, or idea.
- The value of the variable can "vary" from one entity to another.

For example, a person's hair color is a potential variable, which could have the value of "blond" for one person and "brunette" for another.

Discrete Vs. Continuous Variables

Quantitative variables can be further classified as discrete or continuous. If a variable can take on any value between its minimum value and its maximum value, it is called a continuous variable; otherwise, it is called a discrete variable. Some examples will clarify the difference between discrete and continuous variables:

- Suppose the fire department mandates that all fire fighters must weigh between 150 and 250 pounds. The weight of a fire fighter would be an example of a continuous variable; since a fire fighter's weight could take on any value between 150 and 250 pounds.
- Suppose we flip a coin and count the number of heads. The number of heads could be any integer value between 0 and plus infinity. However, it could not be any number between 0 and plus infinity. We could not, for example, get 2.3 heads. Therefore, the number of heads must be a discrete variable.

Univariate Vs. Bivariate Data

Statistical data are often classified according to the number of variables being studied:

- Univariate data: When we conduct a study that looks at only one variable, we say that we are working with univariate data. Suppose, for example, that we conducted a survey to estimate the average weight of high school students. Since we are only working with one variable (weight), we would be working with univariate data.
- Bivariate data: When we conduct a study that examines the relationship between two variables, we are working with bivariate data. Suppose we conducted a study to see if there were a relationship between the height and weight of high school students. Since we are working with two variables (height and weight), we would be working with bivariate data.

Qualitative Variable Vs. Quantitative Variable

Qualitative Variable

A qualitative variable, also called a categorical variable, is a variable that isn't numerical. It describes data that fits into categories. For example:

- Eye colors (variables include: blue, green, brown, hazel).
- States (variables include: Florida, New Jersey, Washington).
- Dog breeds (variables include: Alaskan Malamute, German Shepherd, Siberian Husky, Shih tzu).

These are all qualitative variables as they have no natural order. On the other hand, quantitative variables have a value and they can be added, subtracted, divided or multiplied.

Quantitative Variable

A quantitative variable represents a number. Age of a person, height, marks scored, income, unemployment rate all are the examples of quantitative variables as they are numbers. A quantitative variable is also known as a numeric variable. These variables are in such manner for which some mathematic operations make sense.

There is one general rule for a Quantitative variable that is, a variable is said to be quantitative only if one can add it. For example, marks in English (80), marks in mathematics (85) and marks in science (70) can be added together to make (80+85+70=235), it means that the variables are quantitative.

Quantitative data is useful when it is required to plot the statistical data on a graph. It needs some known units in order to plot it on a graph. A quantitative variable is of two types – discrete or continuous.

- Discrete variables are specific points on a scale. For example, a class might have 30 or 40 students but could not have 30.56 students.
- Continuous variables are continuous on a scale, the values between the figures have meaning and the data can be fragmented into parts. For example, a school mandates that all the classes must have the strength of 30-35 students. So, the strength would be an example of a continuous variable, since the number of students in a class can take any value between 30 and 35 or a, b, c, x, y or z.

Measures of Center

A measure of central tendency is a summary statistic that represents the center point or typical value of a dataset. These measures indicate where most values in a distribution fall and are also referred to as the central location of a distribution. You can think of it as the tendency of data to cluster around a middle value. In statistics, the three most common measures of central tendency are the mean, median, and mode. Each of these measures calculates the location of the central point using a different method.

Mean

Mean is an average of the given numbers: a calculated central value of a set of numbers. In simple words, it is the average. It's also the meanest because it takes the most math to figure it out. Measures of central tendency, or averages, are used in a variety of contexts and form the basis of statistics. The formula to calculate the mean is ratio of the total of observations to number of observations.

Some students get confused with mean and median in statistics. One should always remember that mean is the average value of any given data set while median is the middle value when the data set is arranged in an order. Let us discuss more mean here.

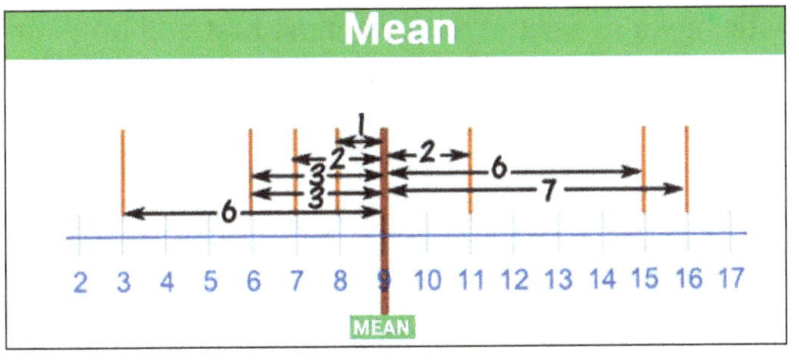

Mean is nothing but the average of given set of values. It denotes the equal distribution of values for a given data set. Central tendency is the statistical measure that recognizes a single value as representative of the entire distribution. It strives to provide an exact description of the whole data. It is the unique value that is the represents collected data. The mean, median and mode are the three commonly used measures of central tendency.

In the case of a discrete probability distribution of a random variable X, the mean is equal to the sum over every possible value weighted by the probability of that value; that is, it is computed by taking the product of each possible value x of X and its probability P(x), and then adding all these products together.

Mean in Maths

Mean, in Maths, is nothing but the average value of the given numbers or data. To calculate the mean, we need to add the total values given in a datasheet and then divide the sum by the total number of values. Suppose, in a data table, the price values of 10 clothing materials are mentioned. If we have to find the mean of the prices, then add the prices of each clothing material and divide the total sum by 10. It will result in average value.

Another example is, if we have to find the average age of students of a class, then we have to add the age of individual student present in the class and then divide the sum by the total number of students present in the class.

Mean is a method which is commonly used in Statistics. In primary schools, we have learned this concept with the term 'average'. But in higher classes, when we are introduced with the topic mean, it refers to an advanced version of sequence or series of a number. In real-world, when there is huge data available, we use statistics to deal with it. Along with mean, we also learn about median and mode. Median is the middle value of a given data when all the values are arranged in ascending order. Whereas mode is the number in the list, which is repeated a maximum number of times.

Mean Formula

The basic formula to calculate the mean is calculated based on the given data set. Each term in the data set is considered while evaluating the mean. The general formula for mean is given by the ratio of sum of all the terms and total number of terms. Hence, we can say:

Mean = Sum of the Given Data/Total number of Data

To calculate the arithmetic mean of a set of data we must first add up (sum) all of the data values (x) and then divide the result by the number of values (n). Since $\sum$ is the symbol used to indicate that values are to be summed we obtain the following formula for the mean ($\bar{x}$):

$$\bar{x}=\sum x/n$$

To find the mean of any given data sets, we have to use take the average. For example, in a class there are 20 students and they have secured a percentage of: 88,82,88,85,84,80,81,82,83,85,84,74,75,76,89,90,89,80,82,83. Find the average of percentage obtained by the class.

Solution: Average = Total of percentage obtained by 20 students in class/Total number of students.

Avg = [88+82+88+85+84+80+81+82+83+85+84+74+75+76+89+90+89+80+82+83]/20

Avg = 1660/20 = 83

Hence, the average of percentage got by each student in class is 83%. In the same, we find the mean of the given data set, in statistics.

Types of Mean

There are majorly three different types of mean value which you will be studying in statistics:

- Arithmetic Mean.
- Geometric Mean.
- Harmonic Mean.

Arithmetic Mean

When you add up all the values and divide by the number of values it is called Arithmetic Mean. To calculate, just add up all the given numbers then divide by how many numbers are given. Example: What is the mean of 3, 5, 9, 5, 7, 2? Now add up all the given numbers:

$3 + 5 + 9 + 5 + 7 + 2 = 31$

Now divide by how many numbers provided in the sequence:

316= 5.16

5.16 is the answer.

Geometric Mean

The geometric mean of two numbers x and y is xy. If you have three numbers x, y, and z, their geometric mean is 3xyz.

Geometric Mean = $\sqrt[n]{x_1 x_2 x_3 \ldots x_n}$

Example: Find the geometric mean of 4 and 3?

Geometric Mean = $\sqrt{4 \times 3} = 2\sqrt{3} = 3.46$.

Harmonic Mean

The harmonic mean is used to average ratios. For two numbers x and y, the harmonic mean is 2xy (x+y). For, three numbers x, y, and z, the harmonic mean is 3xyz (xy+xz+yz),

$$\text{Harmonic Mean(H)} = \frac{n}{\frac{1}{x_1} + \frac{1}{x_2} + \frac{1}{x_3} + \cdots \frac{1}{x_n}}$$

Root Mean Square (Quadratic)

The root mean square, is used in many engineering and statistical applications, especially when there are data points that can be negative.

$$X_{rms} = \sqrt{\frac{x_1^2 + x_2^2 + x_3^2 \ldots x_n^2}{n}}$$

Contraharmonic Mean

The contraharmonic mean of x and y is (x2 + y2)/(x + y). For n values,

$$\frac{x_1^2 + x_2^2 + \ldots + x_n^2}{x_1 + x_2 + \ldots x_n}.$$

Median

The median is the middle number in a sorted, ascending or descending, list of numbers and can be more descriptive of that data set than the average. Median is the middle number in a sorted list of numbers. To determine the median value in a sequence of numbers, the numbers must first be sorted, or arranged, in value order from lowest to highest or highest to lowest. The median can be used to determine an approximate average, or mean, but is not to be confused with the actual mean.

- If there is an odd amount of numbers, the median value is the number that is in the middle, with the same amount of numbers below and above.
- If there is an even amount of numbers in the list, the middle pair must be determined, added together, and divided by two to find the median value.

The median is sometimes used as opposed to the mean when there are outliers in the sequence that might skew the average of the values. The median of a sequence can be less affected by outliers than the mean.

Median Example

To find the median value in a list with an odd amount of numbers, one would find the number that is in the middle with an equal amount of numbers on either side of the median. To find the median, first arrange the numbers in order, usually from lowest to highest.

For example, in a data set of {3, 13, 2, 34, 11, 26, 47}, the sorted order becomes {2, 3, 11, 13, 26, 34, 47}. The median is the number in the middle {2, 3, 11, 13, 26, 34, 47}, which in this instance is 13 since there are three numbers on either side. To find the median value in a list with an even amount of numbers, one must determine the middle pair, add them, and divide by two. Again, arrange the numbers in order from lowest to highest.

For example, in a data set of {3, 13, 2, 34, 11, 17, 27, 47}, the sorted order becomes {2, 3, 11, 13, 17, 27, 34, 47}. The median is the average of the two numbers in the middle {2, 3, 11, 13, 17, 26 34, 47}, which in this case is fifteen {(13 + 17) ÷ 2 = 15}.

Mode

The mode is the value that appears most frequently in a data set. A set of data may have one mode, more than one mode, or no mode at all. Other popular measures of central tendency include the mean, or the average of a set, and the median, the middle value in a set.

In statistics, data can be distributed in various ways. The most often cited distribution is the classic normal (bell-curve) distribution. In this, and some other distributions, the mean (average) value falls at the mid-point, which is also the peak frequency of observed values. For such a distribution, the mean, median, and mode are all the same value. This means that this value is the average value, the middle value, also the mode—the most frequently occurring value in the data.

Mode is most useful as a measure of central tendency when examining categorical data, such as models of cars or flavors of soda, for which a mathematical average median value based on ordering cannot be calculated.

Examples of the Mode

For example, in the following list of numbers, 16 is the mode since it appears more times in the set than any other number:

- 3, 3, 6, 9, 16, 16, 16, 27, 27, 37, 48

A set of numbers can have more than one mode (this is known as bimodal if there are two modes) if there are multiple numbers that occur with equal frequency, and more times than the others in the set.

- 3, 3, 3, 9, 16, 16, 16, 27, 37, 48

In the above example, both the number 3 and the number 16 are modes as they each occur three times and no other number occurs more often. If no number in a set of numbers occurs more than once, that set has no mode:

- 3, 6, 9, 16, 27, 37, 48

A set of numbers with two modes is bimodal, a set of numbers with three modes is trimodal, and any set of numbers with more than one mode is multimodal.

Advantages and Disadvantages of the Mode

Advantages:

- The mode is easy to understand and calculate.
- The mode is not affected by extreme values.
- The mode is easy to identify in a data set and in a discrete frequency distribution.
- The mode is useful for qualitative data.
- The mode can be computed in an open-ended frequency table.
- The mode can be located graphically.

Disadvantages:

- The mode is not defined when there are no repeats in a data set.
- The mode is not based on all values.
- The mode is unstable when the data consist of a small number of values.
- Sometimes data have one mode, more than one mode, or no mode at all.

Measures of Variation

Measures of Variation are amount of variation in a random variable (compare measures of location). Variation is sometimes described as spread or dispersion to distinguish it from systematic trends or differences. Measures of variation are either properties of a probability distribution or sample estimates of them.

Range

In statistics and mathematics, the range is the difference between the maximum and minimum values of a data set and serves as one of two important features of a data set. The formula for a range is the maximum value minus the minimum value in the dataset, which provides statisticians with a better understanding of how varied the data set is.

Two important features of a data set include the center of the data and the spread of the data, and the center can be measured in a number of ways: the most popular of these are the mean, median, mode, and midrange, but in a similar fashion, there are different ways to calculate how spread out the data set is and the easiest and crudest measure of spread is called the range.

The calculation of the range is very straightforward. All we need to do is find the difference between the largest data value in the set and the smallest data value. Stated succinctly we have the following formula: Range = Maximum Value – Minimum Value. For example, the data set 4, 6, 10, 15, 18 has a maximum of 18, a minimum of 4 and a range of 18-4 = 14.

Limitations of Range

The range is a very crude measurement of the spread of data because it is extremely sensitive to outliers, and as a result, there are certain limitations to the utility of a true range of a data set to statisticians because a single data value can greatly affect the value of the range.

For example, consider the set of data 1, 2, 3, 4, 6, 7, 7, 8. The maximum value is 8, the minimum is 1 and the range is 7. Then consider the same set of data, only with the value 100 included. The range now becomes 100 - 1 = 99 wherein the addition of a single extra data point greatly affected the value of the range. The standard deviation is another measure of spread that is less susceptible to outliers, but the drawback is that the calculation of the standard deviation is much more complicated.

The range also tells us nothing about the internal features of the data set. For example, we consider the data set 1, 1, 2, 3, 4, 5, 5, 6, 7, 8, 8, 10 where the range for this data set is 10-1 = 9. If we then compare this to the data set of 1, 1, 1, 2, 9, 9, 9, 10. Here the range is, yet again, nine, however, for this second set and unlike the first set, the data is clustered around the minimum and maximum. Other statistics, such as the first and third quartile, would need to be used to detect some of this internal structure.

Applications of Range

The range is a good way to get a very basic understanding of how spread out numbers in the data set really are because it is easy to calculate as it only requires a basic arithmetic operation, but there are also a few other applications of the range of a data set in statistics.

The range can also be used to estimate another measure of spread, the standard deviation. Rather than go through a fairly complicated formula to find the standard deviation, we can instead use what is called the range rule. The range is fundamental in this calculation.

The range also occurs in a boxplot, or box and whiskers plot. The maximum and minimum values are both graphed at the end of the whiskers of the graph and the total length of the whiskers and box is equal to the range.

Interquartile Range

In Statistics, the interquartile range is the smallest of all the measures of dispersion. It is the difference between the two extreme conclusions of the distribution. In other words, the range is the difference between the maximum and the minimum observation of the distribution. It is defined by:

$$\text{Range} = X_{max} - X_{min}$$

Where X_{max} is the largest observation and X_{min} is the smallest observation of the variable values.

The interquartile range defines the difference between the third and the first quartile. Quartiles are the partitioned values that divide the whole series into 4 equal parts. So, there are 3 quartiles. First Quartile is denoted by Q_1 known as the lower quartile, the second Quartile is denoted by Q_2 and the third Quartile is denoted by Q_3 known as the upper quartile. Therefore, the interquartile range is equal to the upper quartile minus lower quartile.

Interquartile Range Formula

The difference between the upper and lower quartile is known as the interquartile range. The formula for the interquartile range is given below:

Interquartile range = Upper Quartile – Lower Quartile = $Q_3 - Q_1$

where Q_1 is the first quartile and Q_3 is the third quartile of the series.

Semi Interquartile Range

The semi-interquartile range is defined as the measures of dispersion. Semi interquartile range also is defined as half of the interquartile range. It is computed as one half the difference between the 75th percentile (Q_3) and the 25th percentile (Q_1). The semi-interquartile range is one-half of the difference between the first and third quartiles. The Formula for Semi Interquartile Range is:

Semi Interquartile Range = $(Q_3 - Q_1)/2$

Median and Interquartile Range

The median is the middle value of the distribution of the given data. The interquartile range (IQR) is the range of values that resides in the middle of the scores. When a distribution is skewed, and the median is used instead of the mean to show a central tendency, the appropriate measure of variability is the Interquartile range.

- Q_1 – Lower Quartile Part
- Q_2 – Median
- Q_3 – Upper Quartile Part

It is a measure of dispersion based on the lower and upper quartile. Quartile deviation is obtained from interquartile range on dividing by 2, hence also known as semi interquartile range.

Calculate the Interquartile Range

The procedure to calculate the interquartile range is given as follows:

- Arrange the given set of numbers into increasing or decreasing order.
- Then count the given values. If it is odd, then the center value is median otherwise obtain the mean value for two center values. This is known as Q_2 value.
- Median equally cuts the given values into two equal parts. They are described as Q_1 and Q_3 part.
- From Q_1 values, we have to find one median value.
- From Q_3 values, we have to find one median value.
- Finally, we can subtract the median values of Q_1 and Q_3.
- The resulting value is the interquartile range.

Interquartile Range Example

Determine the interquartile range value for the first ten prime numbers.

Given: The first ten prime numbers are:

2, 3, 5, 7, 11, 13, 17, 19, 23, 29

This is already in increasing order. Here the number of values = 10. 10 is an even number. Therefore, the median is mean of 11 and 13. That is,

$Q_2 = 12$

Now we have to get two parts Q_1 and Q_3.

Q_1 part: 2, 3, 5,7,11

Here the number of values = 5. 5 is an odd number. Therefore, the center value is 5, that is,

$Q_1 = 5$

Q_3 part: 13, 17, 19, 23, 29

Here the number of values = 5. 5 is an odd number. Therefore, the center value is 19, that is,

$Q_3 = 19$

The subtraction of Q_1 and Q_3 value is 19 – 5 = 11. Therefore, 11 is the interquartile range value.

Standard Deviation

The standard deviation is a statistic that measures the dispersion of a dataset relative to its mean and is calculated as the square root of the variance. The standard deviation is calculated as the square root of variance by determining each data point's deviation relative to the mean. If the data points are further from the mean, there is a higher deviation within the data set; thus, the more spread out the data, the higher the standard deviation.

Standard deviation is a statistical measurement in finance that, when applied to the annual rate of return of an investment, sheds light on that investment's historical volatility. The greater the standard deviation of securities, the greater the variance between each price and the mean, which shows a larger price range. For example, a volatile stock has a high standard deviation, while the deviation of a stable blue-chip stock is usually rather low. The Formula for Standard Deviation:

$$\text{Standard Deviation} = \sqrt{\frac{\sum_{i=1}^{n}(x_i - \bar{x})^2}{n-1}}$$

Where,

- x_i = Value of the ith point in the data set.
- $\bar{x}$ = The mean value of the data set.
- n = The number of data points in the data set.

Calculating the Standard Deviation

Standard deviation is calculated as follows:

- The mean value is calculated by adding all the data points and dividing by the number of data points.
- The variance for each data point is calculated by subtracting the mean from the value of the data point. Each of those resulting values is then squared and the results summed. The result is then divided by the number of data points less one.
- The square root of the variance—results from above step—is then used to find the standard deviation.

Using the Standard Deviation

Standard deviation is an especially useful tool in investing and trading strategies as it helps measure market and security volatility—and predict performance trends. As it relates to investing, for example, an index fund is likely to have a low standard deviation versus its benchmark index, as the fund's goal is to replicate the index.

On the other hand, one can expect aggressive growth funds to have a high standard deviation from relative stock indices, as their portfolio managers make aggressive bets to generate higher-than-average returns.

A lower standard deviation isn't necessarily preferable. It all depends on the investments and the investor's willingness to assume risk. When dealing with the amount of deviation in their portfolios, investors should consider their tolerance for volatility and their overall investment objectives. More aggressive investors may be comfortable with an investment strategy that opts for vehicles with higher-than-average volatility, while more conservative investors may not.

Standard deviation is one of the key fundamental risk measures that analysts, portfolio managers, advisors use. Investment firms report the standard deviation of their mutual funds and other products. A large dispersion shows how much the return on the fund is deviating from the expected normal returns. Because it is easy to understand, this statistic is regularly reported to the end clients and investors.

References

- Statistics, science: britannica.com, Retrieved 14, May 2020
- Descriptive-statistics: investopedia.com, Retrieved 21, July 2020
- Qualitative-variable: statisticshowto.com, Retrieved 30, April 2020
- Measures-central-tendency-mean-median-mode: statisticsbyjim.com, Retrieved 10, March 2020
- Measures-variation, dictionaries-thesauruses-pictures-and-press-releases, computing: encyclopedia.com, Retrieved 10, August 2020
- What-is-the-range-in-statistics-3126248: thoughtco.com, Retrieved 17, March 2020

2

Understanding Probability

Probability is the branch of mathematics which deals with numerical descriptions of the chances of an event to occur. This chapter explains functions, sequences, limit and continuity, random variables, world of integrals, distribution functions, etc. This chapter has been carefully written to provide an easy understanding of the varied facets of probability.

Probability is the branch of mathematics that studies the possible outcomes of given events together with the outcomes' relative likelihoods and distributions. In common usage, the word "probability" is used to mean the chance that a particular event (or set of events) will occur expressed on a linear scale from 0 (impossibility) to 1 (certainty), also expressed as a percentage between 0 and 100%. The analysis of events governed by probability is called statistics.

There are several competing interpretations of the actual "meaning" of probabilities. Frequentists view probability simply as a measure of the frequency of outcomes (the more conventional interpretation), while Bayesians treat probability more subjectively as a statistical procedure that endeavors to estimate parameters of an underlying distribution based on the observed distribution.

A properly normalized function that assigns a probability "density" to each possible outcome within some interval is called a probability density function (or probability distribution function), and its cumulative value (integral for a continuous distribution or sum for a discrete distribution) is called a distribution function (or cumulative distribution function).

A variate is defined as the set of all random variables that obey a given probabilistic law. It is common practice to denote a variate with a capital letter (most commonly X). The set of all values that X can take is then called the range, denoted R_X. Specific elements in the range of X are called quantiles and denoted x, and the probability that a variate X assumes the element x is denoted P(X=x).

Probabilities are defined to obey certain assumptions, called the probability axioms. Let a sample space contain the union ($\cup$) of all possible events E_i, so,

$$S = (\cup_{(i=1)}^{N} E_i),$$

and let E and F denote subsets of S. Further, let F^'=not-F be the complement of F, so that,

$$F \cup F' = S.$$

Then the set E can be written as:

$$E = E \cap S = E \cap (F \cup F') = (E \cap F) \cup (E \cap F'),$$

where $\cap$ denotes the intersection. Then,

$$P(E)=P(E \cap F)+P(E \cap F')-P[(E \cap F) \cap (E \cap F')],$$

$$=P(E\cap F)+P(E\cap F')-P[(F\cap F')\cap(E\cap E)],$$

$$=P(E\cap F)+P(E\cap F')-P(\phi\cap E),$$

$$=P(E\cap F)+P(E\cap F')-P(\phi),$$

$$=P(E\cap F)+P(E\cap F'),$$

Where φ is the empty set.

Let *P(E|F)* denote the conditional probability of *E* given that *F* has already occurred, then,

$$P(E)=P(E|F)P(F)+P(E|F')P(F')$$
$$=P(E|F)P(F)+P(E|F')[1-P(F)]$$

$$P(A\cap B)=P(A)P(B|A)$$
$$=P(B)P(A|B)$$

$$P(A'\cap B)=P(A')P(B|A')$$

$$P(E|F)=\frac{(E\cap F)P}{P(F)}$$

The relationship:

$$P(A\cap B)=P(A)P(B),$$

holds if A and B are independent events. A very important result states that,

$$P(E\cap F)=P(E)+P(F)-P(E\cap F),$$

which can be generalized to,

$$P(\cup_{i=1}^{\ell} A_i)=$$

$$\sum_i P(A_i)-\sum_{ij}{}' P(A_i\cap A_j)+\sum_{ijk}{}'' P(A_i\cap A_j\cap A_k)-...+(-1)^{n-1}P(\cap_{i=1}^{n} A_i).$$

Functions

A function is a relation between a set of inputs and a set of permissible outputs. Property of the relationship is such that each input is exactly related to one output but two different inputs may result in the same output. We give the mathematical definition as follows:

Let S and T be two subsets of ℝ. A function f from S to T is a subset of S x T such that for each x∈S there exists a unique element $y \in T$ for which $(x,y) \in f$. Here S is called the domain of the function f and the range is a subset of T. The set T is called the co-domain.

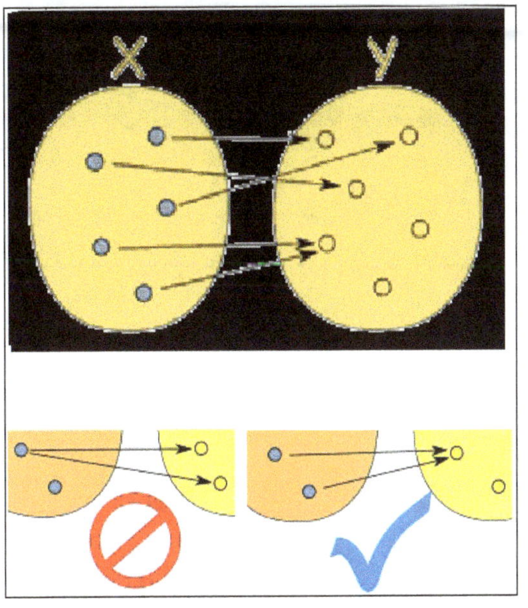

We illustrate further the concept of domain and range through the following examples:

Example:

$$y=f(x)=\frac{1}{x}$$

- The domain set is all real x except x = 0.
- The range set is all real values, f (x) ≠ 0.

Example:

$$y=f(x)=\frac{1}{\sqrt{x-4}}$$

- The domain set is all real x > 4.
- The range set is all real values, f (x) > 0.

Now we have four different types of functions. Assume that the domain set is "A" and the codomain set is "B". A general function points from each member of "A" to a member of "B". When every member of "A" has its own unique matching member in "B" we call the function injective or one-to-one or 1:1. When every member of "B" has at least one (i.e. may be more than one) matching member in "A" we call the function surjective or onto or 1:1. In case the function is both injective and surjective we call it bijective.

A one-to-one (injective) function f from set X to set Y is a function such that each $x \in X$ is related to a different $y \in Y$.

More formally, f:X →Y is 1:1 if either,

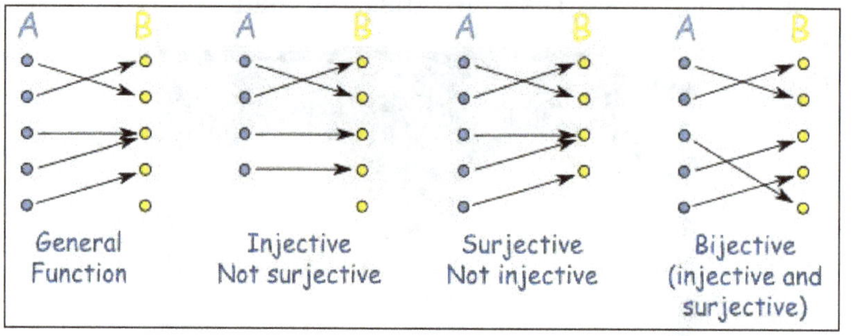

$$f(x_1)=f(x_2) \text{ implies } x_1 = x_2$$

or

$$x_1 \neq x_2 \text{ implies } f(x_1) \neq f(x_2)$$

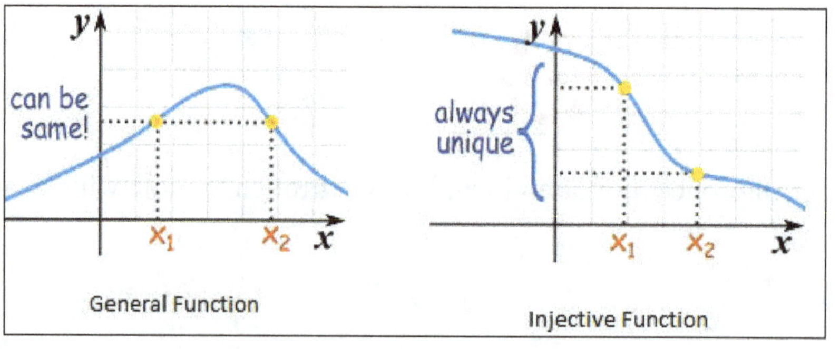

To be injective a horizontal line should never intersect the curve at 2 or more points.

Example: f:ℝ → ℝ is given by f (x) = 3x+7

f (a) = f (b) implies 3a+7 = 3b+7 i.e. a=b.

Example: f:ℝ → ℝ is given by f (x) = x^2

This function is not 1:1 since 16 = f(-4) =f(4).

Sequences

A sequence of real numbers is a function from the set of natural numbers ℕ to the set ℝ of real numbers f: ℕ→ ℝ. If a_n = f (n) for n∈ ℕ then, the sequence is {a_1,a_2,...} or {a_n}. Note that a sequence {a_n} is different from a set {a_n: n ∈ ℕ}. A number may be repeated in a sequence, but it need not be written repeatedly in the set. Instead of sequence of real numbers, a sequence of elements from any non-empty set S, such as sequence of sets, sequence of functions and so on may also be considered. Given a non-empty set S, a sequence in S is a function f: ℕ → S. Consider the following sequences and observe how they vary when n becomes larger and larger.

- a_n = 1; every term of the sequence is same.

- $a_n = n$; terms becomes larger and larger.
- $a_n = \frac{1}{n}$; terms come closer to 0.
- $a_n = \frac{n}{n+1}$; terms come closer to 1.
- $a_n = (-1)^n$; terms of the sequenced oscillates with values 1 and -1.

As n becomes larger and larger, $\{a_n\}$ comes closer to a number, say a. Formally speaking, a sequence $\{a_n\}$ in $\mathbb{R}$ is said to converge to a real number a if for every $\varepsilon > 0$, there exists a positive integer N (in general depending on ε) such that,

$$|a_n - a| < \varepsilon \forall n \geq N.$$

Note that, different ε can result in different N, i.e., the number N may vary as ε varies. When $\{a_n\}$ converges to a, we write,

$$\lim_{n\to\infty} a_n = a.$$

Alternatively,

$$|a_n - a| < \varepsilon \Leftrightarrow a - \varepsilon < a_n < a + \varepsilon.$$

Therefore $\lim_{n\to\infty} a_n = a$ if and only if for every $\varepsilon > 0$, there exists $N \in \mathbb{N}$ such that,

$$a_n \in (a - \varepsilon, a + \varepsilon).$$

Therefore a_n belongs to the open interval $(a-\varepsilon, a+\varepsilon)$ for all N after some finite stage, and this finite stage may vary according as ε varies. Now we give here an important theorem about limit.

Theorem: Limit of a convergent sequence is unique. Suppose $a_n \to a$ and $a_n \to b$ then a=b.

Proof: Suppose $a_n \to a$ and $a_n \to b$ and $a \neq b$. Now for $\varepsilon > 0$, suppose $N_1, N_2 \in \mathbb{N}$ such that,

$$a_n \in (a - \varepsilon, a + \varepsilon) \forall n \geq N_1, a_n \in (b - \varepsilon, b + \varepsilon) \forall n \geq N_2.$$

That is $a_n \in (a - \varepsilon, a + \varepsilon)\ a_n \in (b - \varepsilon, b + \varepsilon) \forall \geq \min(N_1, N_2)$.

Take $\varepsilon < |a - b|/2$, the intervals becomes disjoint. This is a contradiction, hence the proof.

Example: $a_n = \frac{1}{n}$. Then $\lim_{n\to\infty} a_n = 0$. We have to identify $N \in \mathbb{N}$ such that $\frac{1}{n} < \varepsilon$ for all $n \geq N$. It is easy to note that,

$$\frac{1}{n} < \varepsilon \Leftrightarrow n \geq \frac{1}{\varepsilon}$$

$$N = \left\lceil \frac{1}{\varepsilon} \right\rceil + 1.$$

Example: $a_n = \frac{(-1)^n}{n}$, then $\lim_{n\to\infty} a_n = 0$.

Note that $|a_n| = \frac{1}{n}$. Here also, $|a_n - 0| < \varepsilon$ for all $n \geq \left\lceil \frac{1}{\varepsilon} \right\rceil + 1$.

Example: $a_n = 1 - \frac{(-1)^n}{n}$, then $\lim_{n\to\infty} a_n = 1$.

Note here $|a_n - 1| = \frac{1}{n}$. Then we know,

$|a_n - 1| < \varepsilon$ for all $n \geq \left\lceil \frac{1}{\varepsilon} \right\rceil + 1$.

Example: $a_n = c$, constant.

For every $\varepsilon > 0$, $|a_n - c| = 0 < \varepsilon$ for all $n \geq N = 1$.

Example: For a given $k \in \mathbb{N}$, $a_n = \frac{1}{n^k}$, then $\lim_{n\to\infty} a_n = 0$.

Note that $\frac{1}{n^k}, < \varepsilon \Leftrightarrow n \geq \frac{1}{\varepsilon^k}$. Hence, $|a_n - 0| < \varepsilon$ for all $n \geq \left\lceil \frac{1}{\varepsilon} \right\rceil + 1$.

Theorem: Suppose $a_n \to a$ and $b_n \to b$ then,

(a) $a_n \pm b_n \to a \pm b$.

Proof: Since $\lim_{n\to\infty} a_n = a$ and $\lim_{n\to\infty} b_n = b$ for a given $\varepsilon > 0$, however small, we can find two positive integers N_1 and N_2 such that,

$$|a_n - a| < \frac{1}{2}\varepsilon \text{ for } n \geq N_1$$

and

$$|b_n - b| < \frac{1}{2}\varepsilon \text{ for } n \geq N_2$$

Let N be a positive integer which is greater than both N_1 and N_2. Then for every value of $n \geq N$, we have,

$$\begin{aligned}|(a_n \pm b_n) - (a \pm b)| &= |(a_n - a) \pm (b_n - b)| \\ &\leq |(a_n - a) + (b_n - b)| \\ &< \frac{1}{\varepsilon} + \frac{1}{\varepsilon} = \varepsilon\end{aligned}$$

Hence the proof.

(b) For every real number c, $ca_n \to ca$.

Proof: Since $\lim_{n\to\infty} a_n = a$ for a given $\varepsilon > 0$, however small, we can find positive integer N such that,

$$|a_n - a| < \varepsilon \text{ for } n \geq N$$

Hence for any real number C,

$$|ca_n - ca| = |c||a_n - a| < \varepsilon = \varepsilon_1 \text{ (say) for } n \geq N$$

Hence the proof.

(c) If $a_n \leq b_n$ for all $n \in \mathbb{N}$, then $a \leq b$.

Proof: Since $a_n \leq b_n$ for all $n \in \mathbb{N}$ we have $b_n - a_n \geq 0$. From (a) we have $\lim_{n\to\infty} b_n - a_n = b - a \geq 0$. This implies $a \leq b$.

(d) If $a_n \leq c_n \leq b_n$ for all $n \in \mathbb{N}$ and $a = b$, then $c_n \to a$[1].

Proof: Since $\lim_{n\to\infty} a_n = a$ and $\lim_{n\to\infty} b_n = b$ and $a = b$ for a given $\varepsilon > 0$, however small, we can find two positive integers N_1 and N_2 such that,

$$|a_n - a| < \varepsilon \text{ for } n \geq N_1$$

and $|b_n - a| < \varepsilon \text{ for } n \geq N_2$

Let N be a positive integer which is greater than both N_1 and N_2. Then for every value of $n \geq N$, we have,

$$a - \varepsilon < a_n \leq c_n \leq b_n < a + \varepsilon \text{ for } n \geq N$$
$$\Leftrightarrow |c_n - a| < \varepsilon \text{ for all } n \geq N$$
$$\text{i.e.} \lim_{n\to\infty} c_n = a.$$

Example: Find $\lim_{n\to\infty} a_n = \left(1 + \frac{1}{n}\right)^{\frac{1}{n}}$.

Note that $1 \leq a_n \leq \left(1 + \frac{1}{n}\right)$. Using Sandwich theorem $a_n \to 1$.

A sequence which does not converge is called divergent.

A sequence $\{a_n\}$ is said to be divergent and it diverges to $+\infty$, if corresponding to any arbitrarily large positive real number N, there exists a positive integer m such that $a_n > N$, when $n \geq m$. This we express as,

$$\lim_{n\to\infty} a_n = +\infty.$$

A sequence $\{a_n\}$ is said to be divergent and to diverge to $-\infty$, if corresponding to any arbitrarily large positive real number N, there exists a positive integer m such that $a_n < N$, when $n \geq m$. This we express as,

$$\lim_{n\to\infty} a_n = -\infty.$$

If $a_n a_{n+1} < 0 \ \forall \ n \in \mathbb{N}$, then it is an alternating sequence. Alternating sequence may converge ($\{(-1)^n/n\}$) or may diverge $(-1)^n$.

Types of Sequences

Bounded Sequence

A sequence is bounded if it is bounded above and below, that is to say, if there is a number, k, less than or equal to all the terms of sequence and another number, K', greater than or equal to all the terms of the sequence. Therefore, all the terms in the sequence are between k and K'.

$$k \le a_n \le K'$$

Study the following sequences and determine if they are bounded.

$$a_n = \frac{n+2}{2n-1}$$

3, 4/3, 1, 6/7,..

As the sequence is decreasing, 3 is an upper bound and the supremum.

$$a_{1000} = 0.5012506253127.$$

$$a_{1000\,000} = 0.5000012500006.$$

The limit is 0.5.

0.5 is a lower bound and the infimum. Thus, the sequence is bounded.

$$1/2 < a_n \le 3$$

$$a_n = \frac{n}{n+1}$$

$$\frac{1}{2}, \frac{2}{3}, \frac{3}{4}, \frac{4}{5}, \frac{5}{6}, \ldots$$

As the sequence is increasing, 1/2 is a lower bound and the infimum.

$$a_{1000} = 0.999000999001.$$

$$a_{1000\,000} = 0.999999000001.$$

The limit is 1. 1 is an upper bound and the supremum. Thus, the sequence is bounded.

$$0.5 \le a_n < 1.$$

Subsequence

A subsequence of a sequence $\{a_n\}_{n=0}^{\infty}$ is a sequence formed by deleting elements of the a_n to produce a new a_n. This subsequence is usually written as a_{nk}, where $\{n_k\}_{k=0}^{\infty}$ is an increasing sequence of positive integers.

Because all the terms of a subsequence are also terms of the original sequence, the properties of the subsequence are closely tied to the properties of the sequence, and so a great deal of information about one can be determined by studying the other. Sequences are objects of central importance in analysis, but they can be difficult to work with, and their simpler subsequences provide a useful tool for understanding them.

Convergence of Subsequences

Theorem: A sequence converges to a limit x if and only if every subsequence also converges to the limit x.

Proof: For one direction, suppose that $a_n \to x$, and consider some subsequence a_{nk}. Because $a_n \to x$, for every $\epsilon > 0$, there is $N > 0$ such that,

$$n > N \Rightarrow |a_n - x| < \varepsilon$$

Then, for a_{nk}, there must be some K such that $k > K \Rightarrow n_k > N$ implies, because the n_k are increasing. But if $n_k > N$, then $|a_{nk} - x| < \varepsilon$, so $a_{nk} \to x$ as well. Thus any subsequence converges as well.

For the other direction, suppose that every subsequence of a_{nk} converges, and suppose for the sake of contradiction that a_n does not converge. Then, for every $I > 0$, there is an arbitrarily large a_n such that $|a_n - x| > \frac{1}{i}$ (if this were not true, then the sequence would converge). Then, taking the a_n that satisfy this for each i, choosing them to be increasing, gives a subsequence that does not converge; contradiction. This fact can be extremely useful in characterizing the convergence or divergence of sequences.

Example: Does the sequence $a_n = \frac{1}{3n+1}$ converge? Yes. This is a subsequence of the convergent sequence $a_n = \frac{1}{n}$, so it must converge.

Bolzano-Weierstrass

A very important theorem regarding the convergence of subsequences of sequences in $\mathbb{R}$ is the Bolzano-Weierstrass theorem.

Theorem: Every bounded sequence in $\mathbb{R}$ has a convergent subsequence.

Proof: We claim that every sequence in $\mathbb{R}$ has a monotone subsequence. Because every bounded monotone sequence converges, the result follows. Consider a sequence a_n, and call a term a_k a peak if it is larger than every subsequent term in the sequence. Now, if a sequence has infinitely many peaks, the subsequence corresponding to the peaks is monotonically decreasing. If there are only finitely many peaks, suppose that the last peak is at N. Then, the N+1 term is not a peak, so there is a term after it larger than it. That term is also not a peak, so there is a term after that larger than it. In this way, a monotonically increasing subsequence can be constructed. Either way, the sequence has a monotone subsequence, and this subsequence will converge, proving the result.

The Bolzano-Weierstrass theorem can be used to give proofs of the Ascoli-Arzela theorem, as well as be applied to various problems in economics.

Cauchy Sequence

A sequence $\{a_n\}$ in a metric space X is said to be a Cauchy sequence if for every ε (>0), there exists an integer n_0 such that $|a_n - a_m| < \varepsilon$ if $n > n_0$ and $m > n_0$. A necessary and sufficient condition for a real sequence $\{a_n\}$ to be convergent is that it is a Cauchy sequence. Sometimes it is easier in terms of Cauchy sequence to evaluate the property whether a sequence is convergent or not. Note: A metric space X in which every Cauchy sequence converges is said to be complete.

Cauchy sequences are useful because they give rise to the notion of a complete field, which is a field in which every Cauchy sequence converges. Because the Cauchy sequences are the sequences whose terms grow close together, the fields where all Cauchy sequences converge are the fields that are not 'missing' any numbers. The canonical complete field is $\mathbb{R}$, so understanding Cauchy sequences is essential to understanding the properties and structure of $\mathbb{R}$.

Cauchy Sequences in $\mathbb{R}$

The definition of Cauchy sequences given above can be used to identify sequences as Cauchy sequences. Example: Is the sequence $a_n = \frac{1}{2^n}$ a Cauchy sequence? Yes. Take any $\epsilon > 0$, choose N so that large $2^{-N} < \epsilon$. Then, if n, m > N, we have,

$$|a_n - a_m| = \left|\frac{1}{2^n} - \frac{1}{2^m}\right| \leq \frac{1}{2^n} + \frac{1}{2^m} \leq \frac{1}{2^N} + \frac{1}{2^N} = \varepsilon$$

so this sequence is Cauchy.

Showing that a sequence is not Cauchy is slightly trickier. For a sequence not to be Cauchy, there needs to be some N > 0 such that for any $\epsilon > 0$, there are m, n > N with $|a_n - a_m| > \varepsilon$. In other words, no matter how far out into the sequence the terms are, there is no guarantee they will be close together.

Example: Is the sequence $a_n = n$ a Cauchy sequence? No. Take $\epsilon=1$. Then, for any N, if we take n = N +3 and m = N+1, we have that $|a_m - a_n| = 2 > 1$, so there is never any N that works for this ϵ. Thus, the sequence is not Cauchy. When attempting to determine whether or not a sequence is Cauchy, it is easiest to use the intuition of the terms growing close together to decide whether or not it is, and then prove it using the definition.

Cauchy Sequences in an Abstract Metric Space

The ideas from the previous sections can be used to consider Cauchy sequences in a general metric space (X, d). In this context, a sequence $\{a_n\}$ is said to be Cauchy if, for every $\epsilon > 0$, there exists N > 0 such that,

$$M, n > n \Rightarrow d(a_m, a_n) < \epsilon.$$

On an intuitive level, nothing has changed except the notion of "distance" being used. Example: Consider the metric space of continuous functions on [0,1] with the metric,

$$d(f,g) = \int_0^1 |f(x) - g(x)|\,dx.$$

Is the sequence $f_n(x) = nx$, a Cauchy sequence in this space? Note that,

$$d(f_m, f_n) = \int_0^1 |mx - nx|\,dx = \left[|m-n|\frac{x^2}{2}\right]_0^1 = \frac{|m-n|}{2}.$$

By taking m = n+1, we can always make this $\frac{1}{2}$, so there are always terms at least $\frac{1}{2}$ apart, and thus this sequence is not Cauchy.

Cauchy Sequences and Convergence

Cauchy sequences are intimately tied up with convergent sequences. For example, every convergent sequence is Cauchy, because if $a_n \to x$, then,

$$|a_m - a_n| \le |a_m - x| + |x - a_n|,$$

both of which must go to zero. The converse of this question, whether every Cauchy sequence is convergent, gives rise to the following definition: A field is complete if every Cauchy sequence in the field converges to an element of the field.

Example: Is $\mathbb{Q}$ a complete field? No. Take a sequence given by $a_0 = 1$ and satisfying $a_n = \frac{a_{n-1}}{2} + \frac{1}{a_n}$.

This sequence has limit $\sqrt{2}$, so it is Cauchy, but this limit is not in $\mathbb{Q}$,so $\mathbb{Q}$ is not a complete field. In particular, $\mathbb{R}$ is a complete field, and this fact forms the basis for much of real analysis: to show a sequence of real numbers converges, one only need show that it is Cauchy. Similarly, given a Cauchy sequence, it automatically has a limit, a fact that is widely applicable.

Monotonic Sequence

A sequence (a_n) is monotonic increasing if a_n+1≥ an for all n ∈ N. The sequence is strictly monotonic increasing if we have > in the definition. Monotonic decreasing sequences are defined similarly.

Theorem: A bounded monotonic increasing sequence is convergent.

Proof: We will prove that the sequence converges to its least upper bound (whose existence is guaranteed by the Completeness axiom). So let α be the least upper bound of the sequence. Given ε > 0, we'll show that all the terms of the sequence (except the first few) are in the interval (α - ε, α + ε).

Now since α + ε is an upper bound of the sequence, all the terms certainly satisfy $a_n < \alpha + \varepsilon$. Also since α - ε is not an upper bound of the sequence, we must have $a_N > \alpha - \varepsilon$ for some N. But then all the later terms will be > α - ε also and so (for n > this N) we have our condition for convergence.

A similar result is true for a bounded monotonic decreasing sequence (which converges to its greatest lower bound). There are sequences which are convergent without being monotonic. For example, the sequences (-1, 1/2, -1/3, 1/4,...) and (1/2, $1/2^2$, 1/3, $1/3^2$,...) both converge to 0. For sequences given by recurrence relations it is sometimes easy to see what their limits are.

Example: Define a sequence by $a_{n+1} = \sqrt{(2a_n - 1)}$ with $a_1 = 2$. Then suppose that this has a limit α. Then for large n, we have $a_n = \alpha$ (approx) and $a_{n+1} = \alpha$ (approx) and so we must have $\alpha = \sqrt{(2\alpha - 1)}$ and hence $\alpha^2 = 2\alpha - 1$ and we get α = 1.

To show that it does indeed have a limit, we'll prove that it is monotonic decreasing and bounded below. Since the terms of the sequence are positive, the sequence is clearly bounded below by 0. To show monotonicity:

$$a_{n+1} - a_n = \sqrt{(2a_n - 1)} - \sqrt{(2a_{n-1} - 1)}$$

and if we now use the fact that $\sqrt{x} - \sqrt{y} = (x - y)/(\sqrt{x} + \sqrt{y})$. We get $a_{n+1} - a_n = 2(a_n - a_n - 1) / (\sqrt{(2a_n - 1)} + \sqrt{(2a_{n-1} - 1)}$ and the result then follows by induction.

Define $a_n+1 = (a_n^2 + 1)/_2$ with $a_1 = 2$.

Claim: This is monotonic increasing.

$a_{n+1} - a_n = (a_n^2 - a_{n-1}^2)/_2 = (a_n + a_{n-1})(a_n - a_{n-1})/_2$ and so (since $a_n > 0$) the result follows by induction. What is the limit?

The limit α satisfies $\alpha = (\alpha^2 + 1)/_2$ and hence $\alpha = 1$. But since the sequence starts at 2 and increases, it cannot converge to 1 and hence it has no limit. One may show that if one takes the starting value $a_1 = 1/2$ then the sequence is bounded above and does converge to 1.

Lower and Upper Limits

For every real M there exists an integer N such that $n \geq N$ implies $a_n \geq M$. We write $a_n \to \infty$. For every real M there exists an integer N such that $n \geq N$ implies $a_n \leq M$. We write $a_n \to -\infty$. Let E be the set of numbers x (in extended real number system) such that $a_{nk} \to x$ for some subsequence $\{a_{nk}\}$. Thus E contains all the sub-sequential limits. If we put:

$$a^* = \sup E \text{ and } a_* = \inf E,$$

The numbers a^* and a_* are called upper and lower limits of $\{a_n\}$. The notation for a^* and a_* are as follows:

$$\limsup_{n\to\infty} a_n = a^*, \liminf_{n\to\infty} a_n = a_*,$$

Example: $\{a_n\}$: A sequence containing all rational numbers. Here every real number is a sub-sequential limit.

$$\limsup_{n\to\infty} a_n = \infty, \liminf_{n\to\infty} a_n = \infty,$$

Example: $a_n = (-1)^n (1 + \frac{1}{n})$: $\limsup_{n\to\infty} a_n = 1$, $\liminf_{n\to\infty} a_n = 1$.

Example: if $\lim_{n\to\infty} a_n = a$ and only if $\limsup_{n\to\infty} a_n = \liminf_{n\to\infty} a_n = a$.

Three Sequences

1. For $p > 0$, $\lim_{n\to\infty} \frac{1}{n^P} = 0$.

Proof: Since $\left|\frac{1}{n^p} - 0\right| < \varepsilon$ whenever $n > (1/\varepsilon)^{1/p}$. Hence by the definition of limit the proof is complete.

2. For $p > 0, \lim_{n\to\infty} p^{1/n} = 1$.

Proof: Take $p > 1$. Put $x_n = p^{1/n} - 1$. Then $x_n > 0$. Using binomial theorem,

$$p=(1+x_n)^n > 1+nx_n \Rightarrow 0 < x_n < \frac{p-1}{n}.$$

Therefore by Sandwich theorem, $x_n \to 0$ as $n \to \infty$.

- For $p = 1$, the proof is trivial.
- For $p < 1$, the proof follows by taking reciprocals.

3. $\lim_{n\to\infty} n^{1/n} = 1$.

Proof: Put $x_n = n^{1/n} - 1$. Then $x_n > 0$. Using binomial theorem,

$$n = (1+x_n)^n \geq \frac{n(n-1)}{2} x_n^2.$$

$\therefore 0 \leq x_n \leq \sqrt{\frac{2}{n-1}}$ [Assuming that $n \geq 2$].

$$x_n \to 0 \text{ as} n \to \infty.$$

Limit and Continuity

The limit of a function f(x) as x approaches the value a from both sides (i.e. from both bigger and smaller values), is that quantity l, (if one such exists), towards which f(x) approaches, and from which the numerical difference of f(x) can be made as small as we please by making x approach a sufficiently closely. It is called the Limit of f(x) as x tends to a and is written as $\lim_{n\to\infty} f(x) = l$.

Let f be a real valued function defined in some open interval E containing a point a, although we do not insist that f be defined on the point 'a' itself and let l be a real number. We write $f(x) \to l$ as $x \to a$, meaning that, for every $\varepsilon > 0$ there exists a $\delta > 0$ such that,

$$|f(x) - l| > \varepsilon \text{ whenever } |x - a| < \delta.$$

We denote this limit by $\lim_{n\to\infty} f(x) = l$.

- Instead of defining on a subset of real line, we can always define limit on general metric spaces with appropriate metric.
- f may not necessarily be defined at the point a itself.
- Even if f is defined at x = a, we may very well have $\lim_{x\to a} f(x) \neq f(a)$.
- If f has a limit x = a, this limit is unique.

Theorem: Let $E \subset \Re$, a is a limit point of E, f and g are defined on E. Then,

$$\lim_{x\to a}(f(x)+g(x))=\lim_{x\to a}f(x)+\lim_{x\to a}g(x).$$

Proof: Suppose $\lim_{x\to a} f(x)=l$ and $\lim_{x\to a} g(x)=m$ where l and m are finite quantities. Since $\lim_{x\to a} f(x)=l$ and $\lim_{n\to\infty} g(x)=m$, for any given positive number ε, we can choose positive numbers δ_1, δ_2 such that,

$$|f(x)-l|<\frac{1}{2}\varepsilon \text{ when } 0<|x-a|<\delta_1,$$

$$|g(x)-m|<\frac{1}{2}\varepsilon \text{ when } 0<|x-a|<\delta_2,$$

Let δ be any positive number which is smaller than both δ_1 and δ_2; then the inequalities above equations both hold good when $0 < |x - a| < \delta$. ∴ By definition, $l + m$ is the limit of $f(x) + g(x)$ as $x \to a$.

$$\lim_{x\to a}(f(x)\cdot g(x))=\lim_{x\to a}f(x)\cdot\lim_{x\to a}g(x).$$

Proof: We can write, $f(x)\,g(x) - lm = \{f(x) - l\}\{g(x) - m\} + m\{f(x) - l\} + l\{g(x) - m\}$. ∴ $|f(x)\,g(x) - lm| \le |f(x) - l||g(x) - m| + |m||f(x) - l| + |l||g(x) - m|$. Now ε being any pre-assigned positive quantity, and choosing any other positive quantity k,

$$k, \frac{\varepsilon}{3|l|}, \frac{\varepsilon}{3|m|}, \frac{\varepsilon}{3|k|}$$

are known positive quantities. Therefore choosing positive numbers δ_1, δ_2, δ_3, δ_4 such that,

$$|f(x)-l|<k \text{ when } 0<|x-a|<\delta_1,$$

$$|g(x)-m|<\frac{\varepsilon}{3k} \text{ when } 0<|x-a|<\delta_2,$$

$$|f(x)-l|<\frac{\varepsilon}{3|m|} \text{ when } 0<|x-a|<\delta_3,$$

$$|g(x)-m|<\frac{\varepsilon}{3|l|} \text{ when } 0<|x-a|<\delta_4,$$

and proceeding as before we can show that $\lim_{x\to a}(f(x)\cdot g(x))=l\cdot m$.

$$\lim_{x\to a}(f(x)/g(x))=\lim_{x\to a}f(x)/\lim_{x\to a}g(x) \text{ provided } g(x)\neq 0.$$

Proof: We can write $\left|\frac{f(x)}{g(x)}-\frac{1}{m}\right|=\left|\frac{m\{f(x)-l\}-l\{g(x)-m\}}{|m||g(x)|}\right|\le\left|\frac{|m||f(x)-l|-|l|g(x)-m|}{|m||g(x)|}\right|$. Now since

$\lim_{x\to a} g(x)=m$, there exists a positive number δ_1 such that $|g(x) - m|\ \frac{1}{2}|m|$ when $0 < |x - a| < \delta_1$, for $m \neq 0$.

$$\therefore |m|-|g(x)|\le|g(x)-m|<\frac{1}{2}|m|$$

Or,

$$|g(x)| > \frac{1}{2}|m| \text{ when } 0 < |x-a| < \delta_1.$$

Hence proceeding as before we can show that $\lim_{x\to a}(f(x)/g(x)) = \lim_{x\to a} f(x)/\lim_{x\to a} g(x)$.

One Sided Limits

Let f be a real valued function defined in some open interval E containing a point a (while f may not be defined on the point a itself) and let l_1 be a real number. We write $f(x) \to l_1$ as $x \to a+$, meaning that, for every $\varepsilon > 0$ there exists a $\delta > 0$ such that,

$$|f(x) - l_1| < \varepsilon \text{ whenever } 0 < x - a < \delta.$$

Thus the limit of a function f(x) as x approaches the value a from the right (i.e. from bigger values), is that quantity l_1, (if one such exists), towards which f(x) approaches, and from which the numerical difference of f(x) can be made as small as we please by making x approach a sufficiently closely, all the time keeping it greater than a. It is called the Right-hand limit of f(x) as x tends to a and is written as $\lim_{x\to a+} f(x) = l_1$.

Let f be a real valued function defined in some open interval E containing a point a (while f may not be defined on the point a itself) and let l_2 be a real number. We write $f(x) \to l_2$ as $x \to a-$, meaning that, for every $\varepsilon > 0$ there exists a $\delta > 0$ such that,

$$|f(x) - l_1| < \varepsilon \text{ whenever } 0 < a - x < \delta.$$

Thus the limit of a function f(x) as x approaches the value a from the left (i.e. from smaller values), is that quantity l_2, (if one such exists), towards which f(x) approaches, and from which the numerical difference of f(x) can be made as small as we please by making x approach a sufficiently closely, all the time keeping it smaller than a. It is called the Left-hand limit of f(x) as x tends to a and is written as $\lim_{x\to a-} f(x) = l_2$. We denote this limit by $\lim_{x\to a+} f(x) = l_1$.

- If $l_1 = l_2$, we say that limit exists.
- Even if $l_1 = l_2 = l$(say), then l may not necessarily be equal to f(a).

Example: Suppose, $f(x) = x^2 + 3$, $x \in \Re$. To check whether limit exists at x = a. Here $\lim_{x\to 0+} f(x) = 3$, $\lim_{x\to 0-} f(x) = 3$. In fact, f(0) = 3.

Example: Suppose, $f(x) = \begin{cases} x^2 + 3, & \text{if } x \neq 0 \\ 4 & \text{if } x = 0 \end{cases}$

Limit exists at x = 0, because, $\lim_{x\to 0+} f(x) = 3$, $\lim_{x\to 0-} f(x) = 3$. Although f(0) ≠3.

Example: Suppose, $f(x) = \begin{cases} x^2 + 3, & \text{if } x \geq 0 \\ x^2 + 4 & \text{if } x = 0 \end{cases}$

Limit does not exists at x = 0, because $\lim_{x\to 0+} f(x) = 3$, $\lim_{x\to 0-} f(x) = 4$.

Continuity

In mathematics, continuity is rigorous formulation of the intuitive concept of a function that varies with no abrupt breaks or jumps. A function is a relationship in which every value of an independent variable—say x—is associated with a value of a dependent variable—say y. Continuity of a function is sometimes expressed by saying that if the x-values are close together, then the y-values of the function will also be close.

For close x-values, the distance between the y-values can be large even if the function has no sudden jumps. For example, if y = 1,000x, then two values of x that differ by 0.01 will have corresponding y-values differing by 10. On the other hand, for any point x, points can be selected close enough to it so that the y-values of this function will be as close as desired, simply by choosing the x-values to be closer than 0.001 times the desired closeness of the y-values. Thus, continuity is defined precisely by saying that a function f(x) is continuous at a point x_0 of its domain if and only if, for any degree of closeness ε desired for the y-values, there is a distance δ for the x-values (in the above example equal to 0.001ε) such that for any x of the domain within the distance δ from x_0, f(x) will be within the distance ε from $f(x_0)$. In contrast, the function that equals 0 for x less than or equal to 1 and that equals 2 for x larger than 1 is not continuous at the point x = 1, because the difference between the value of the function at 1 and at any point ever so slightly greater than 1 is never less than 2.

A function is said to be continuous if and only if it is continuous at every point of its domain. A function is said to be continuous on an interval, or subset of its domain, if and only if it is continuous at each point of the interval. The sum, difference, and product of continuous functions with the same domain are also continuous, as is the quotient, except at points at which the denominator is zero. Continuity can also be defined in terms of limits by saying that f(x) is continuous at x_0 of its domain if and only if, for values of x in its domain,

$$\lim_{x \to x_0} f(x) = f(x_0).$$

A more abstract definition of continuity can be given in terms of sets, as is done in topology, by saying that for any open set of y-values, the corresponding set of x-values is also open. (A set is "open" if each of its elements has a "neighborhood," or region enclosing it, that lies entirely within the set.) Continuous functions are the most basic and widely studied class of functions in mathematical analysis, as well as the most commonly occurring ones in physical situations.

Random Variables

Statistics is all about randomness. So the one concept which lies at the heart of statistics is random variable. As the name suggests 'random' and 'variable', here 'variable' means it will vary, it will take some values, it will belong to some intervals with some law or rule that is governed by this process 'random'.

In many cases we shall find the elements of Ω, i.e. the elementary events are themselves expressed as real numbers. For example while throwing a die once any of the numbers 1, 2,..., 6 can appear.

Therefore the sample space is $\Omega = \{1, 2,..., 6\}$. Even when the elementary events themselves are not numbered we can associate some definite real number to each elementary event according to some given rule. In tossing a coin once the sample space Ω is {H,T}. If we assign the number 1 to H and 0 to T then the sample space can also be written as $\Omega = \{0, 1\}$. Often when an experiment is performed our main interest lies in the function of the outcome rather than the actual outcome itself. For example, in a dice throwing game a gambler is more interested in his gains rather than his number of wins and losses. It is therefore desirable to introduce a point function on the sample space. These quantities of interest or more formally these real valued functions defined on the sample space are known as random variables. We give here the formal definition of random variable:

By a random variable with respect to the probability space $(\Omega,F,\mathbb{P})$ we mean a finite real valued function, say X such that,

$$\{\omega:X(\omega)\le x\}\in\mathscr{F}\ \forall x\in\mathbb{R}$$

Example: A coin is tossed twice. Denoting the occurrence of 'head' and 'tail' by 'H' and 'T' respectively we have here the sample space Ω as:

$$\Omega\{HH,HT,TH,TT\}=\{\omega_1,\omega_2,\omega_3,\omega_4\},$$

F = class of all subsets of Ω.

Define $X(\omega)$ = number of heads. Then,

$$X(\omega)=\begin{cases}0\text{ if }\omega=\omega_4\\1\text{ if }\omega=\omega_2\text{ or }\omega_3\\2\text{ if }\omega=\omega_1\end{cases}$$

$$\{\omega:X(\omega)\le x\}=\begin{cases}\phi\in\mathscr{F} & \text{if } x<0\\\{TT\}\in\mathscr{F} & \text{if } x\le 0<1\\\{TH,HT,TT,HH\}\in\mathscr{F} & \text{if } 1\le x<2\\\{TH,HT,TT,HH\}\in\mathscr{F} & \text{if } x\ge 2\end{cases}$$

So by definition of random variable we can say that the number of heads is a random variable since it satisfies all the conditions that is mentioned above.

A real valued function X defined on the probability space $(\Omega,F,\mathbb{P})$ is called a random variable if,

$$X^{-1}(B)=\{\omega:X(\omega)\in B\}\in\mathscr{F}\ \forall B\in\mathscr{B}$$

where $\mathscr{B}$ = Borel, σ = field.

This is the definition where we use the concept of Borel σ-field. Following are equivalent statements: X is a random variable if and only if,

(i) $\{\omega:X(\omega)>x\}\in\mathscr{F}\ \forall x\in\mathbb{R}$

(ii) $\{\omega : X(\omega) \geq x\} \in \mathscr{F}\ \forall x \in \mathbb{R}$

(iii) $\{\omega : X(\omega) < x\} \in \mathscr{F}\ \forall x \in \mathbb{R}$

(iv) $\{\omega : a < x(\omega) < b\} \in \mathscr{F}\ \forall x \in \mathbb{R}, -\infty < a \leq b < \infty$

Examples: Random Variable or Not

Example: Suppose A be an arbitrary event in Ω. Then $\mathbb{1}_A$ is a random variable if and only if $A \in \mathscr{F}$ Since,

$$\{\omega : X(\omega) \leq x\} = \begin{cases} \phi & \text{if } x < 0 \\ A^c & \text{if } 0 \leq x < 1 \\ \Omega & \text{if } x \geq 1 \end{cases}$$

Example: Suppose a fair dice is rolled once. Here the sample space is: Ω = {1, 2, 3, 4, 5, 6}.

$X : \Omega \to \mathbb{R}$ be such that: X(1) = X(3) = X(5) = 0 and X(2) = X(4) = X(6) = 1. Here X is a r.v. if and only if {1, 3, 5} $\in \mathscr{F}$ Since,

$$\{\omega : X(\omega) \leq x\} = \begin{cases} \phi & \text{if } x < 0 \\ \{1,3,5\} & \text{if } 0 \leq x < 1 \\ \Omega & \text{if } x \geq 1 \end{cases}$$

Some Results

Now that we have defined a random variable, for example X is a random variable, the obvious question that comes to our mind is: What will happen to any function of X i.e. f (X)? Whether f (X) is also a random variable? If f be any arbitrary function will it be a random variable? If not then what sort of condition we require so that f (X) is also a random variable? Our next theorems are going to shed some light upon such questions.

Theorem: Let X be a random variable. Then,

(a) X^2 is a random variable.

Proof: $\{\omega : X^2(\omega) \leq x\} = \begin{cases} \phi \in \mathscr{F} & \text{if } x < 0 \\ \{\omega : X(\omega) - 0\} \in \mathscr{F} & \text{if } x - 0 \\ \{\omega : -\sqrt{x} - X(\omega) < \sqrt{x}\} \in \mathscr{F} & x > 0 \end{cases}$

(b) Let f (X) be any continuous function of X. Then f (X) is a random variable:

Proof: The function f is continuous if and only if for every open set G in $\mathbb{R}$, the set f^{-1} (G) is open. So let us begin with an arbitrary open set G in $\mathbb{R}$.

$$\{\omega : f(X(\omega)) \in G\}$$
$$= \{\omega : X(\omega) \in f^{-1}(G) \in \mathscr{F}\}$$

Since f is continuous, G is open $\Leftrightarrow f^{-1}(G)$ is open $\Rightarrow f^{-1}(G) \in \mathscr{B}$ since any open set is a Borel set.

Theorem: Let Y and Z be two random variables. Then,

(a) Y + Z (b) cY (c) Y Z (d) max(Y,Z) (e) min(Y,Z) (f) $X^+ = \max(X, 0)$ (g) $X^- = -\min(X, 0)$ are also random variable.

Proof: (a) Note that Y + Z > x if and only if there is a rational number q in the interval Y > q > x - Z so that Y > q and Z > x - q. i.e.

$$[Y+Z>x] = \bigcup_{q\in\mathbb{Q}} \underbrace{[Y>q]}_{\in\mathscr{F}} \cap \underbrace{[Z>x-q]}_{\in\mathscr{F}}$$

$$\therefore [Y+Z>x] \in \mathscr{F}$$

(b) In case c > 0, $[cY < x] = [Y > \frac{x}{c}] \in \mathscr{F}$. In case c < 0, $[cY < x] = [Y > \frac{x}{c}] \in \mathscr{F}$

(c) $YZ = \frac{1}{2}[(Y+Z)^2 - Y^2 - Z^2]$. Now, $(Y + Z)^2, Y^2, Z^2$ are all random variables. Therefore using (a) and (b) part of above theorem we see that Y - Z is also a random variable.

(d) We can write $\max(Y,Z) = \frac{1}{2}(|Y - Z| + Y + Z)$. Since $|Y - Z|$ is a continuous function of Y - Z it is a r.v. Hence the result follows using (a) and (b) of above theorem.

(e) We can write $\min(Y,Z) = \frac{1}{2}(|Y - Z| - (Y+ Z))$. The proof of this result is now similar to part (d).

(f) The proof of this part is similar to that of part (d).

(g) The proof of this part is similar to that of part (e).

Theorem: Let $\{X_n\}$ be a sequence of random variables. Then,

(a) Inf X_n (b) lim inf X_n (c) sup X_n (d) lim sup X_n are also random variables.

Proof: In order to prove part (a) we can write $\inf X_n \geq x = \bigcap_{n=1}^{\infty}[X_n \geq x]$. In order to prove part (b) we can write $\liminf X_n \geq x = [X_n \geq x - 1/m$ for all but finite n] for all m = 1, 2,.... For the last two cases replace X_n by $-X_n$.

Transformation

Let X be random variable on $(\Omega, \mathscr{F}, \mathbb{P})$ and $f : \mathbb{R} \to \mathbb{R}$ be a function. The question that one will eventually ask: Is f (X) a random variable? We give the answer as follow:

If f is arbitrary, f(X) may not be a random variable. If f is Borel-measurable, that is pre-images of Borel sets under f are also Borel sets then and only then f(X) is a random variable. It is clear that the pre-images of Borel sets under the composite function $f \circ X(.)$ are events.

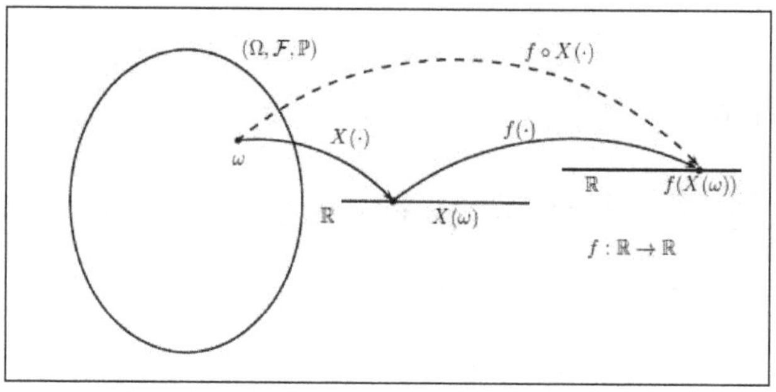

Then f (X) is a random variable. We can extend this concept from univariate to multivariate setup. For a Borel measurable function $f: \mathbb{R}^n \rightarrow \mathbb{R}$ and random variables $X_1, X_2, ..., X_n$, $f(X_1, X_2, ..., X_n)$ is also a random variable.

Random Experiment

The concept of random variable has high correspondence to random experiment. A random experiment is a process characterized by the following properties:

(a) It is performed according to some set of rules.

(b) It can be repeated arbitrarily often.

(c) The result of each performance depends on chance and cannot be predicted uniquely.

Example: Tossing of a coin–The outcome of a trial can be either head or tail showing up. So we don't know which one will come whenever we are tossing a coin.

Example: Whenever we are throwing a die we do not know which one will come but we know that one of the numbers {1, 2, 3, 4, 5, 6} will come.

Sequential Random Experiments: By sequential random experiment we mean performing a sequence of simple random sub-experiments e.g. first toss a coin, then throw a dice. Sometimes, the second sub-experiment depends on the outcome of the first; e.g. throw a dice first, if it is even, then toss a coin. A random experiment may involve a continuum of measurements say, the height of a student which takes some value between 165 cm to 170cm.

Sample Space

Whatever be the random variable corresponding to the random variable we will have sample space. Sample space Ω of a random experiment is defined as the set of all possible outcomes. These Outcomes are mutually exclusive in the sense that they cannot occur simultaneously. A sample space can be finite, countable infinite or uncountable infinite. We give below some examples of sample space.

- Toss a coin two times Ω_1 = {(H,H), (H,T), (T,H), (T,T)}. Ω_1 is countable. Ω_1 is called a discrete sample space. Define B = {H,T}, then $\Omega_1 = B \times B$.
- Toss a dice until a 'six' appears and count the number of times the dice was tossed. Ω_2 = {1, 2, 3,...} is discrete and countable infinite (one-to-one correspondence with positive integers).

- Pick a number X at random between zero and one, then pick a number Y at random between zero and X. $\Omega 3 = \{(x, y): 0 \le y \le x \le 1\}$ is a continuous (uncountable infinite) sample space.

Coin Example

Suppose a coin is tossed twice. Here the sample space is $\Omega = \{HH, HT, TH, TT\}$ and $\mathscr{F} = \sigma(X)$. Define X = Number of heads.

$$X(\omega)=\begin{cases}0 & \text{if } \omega=\omega_4\\ 1 & \text{if } \omega=\omega_2 \text{ or } \omega_3\\ 2 & \text{if } \omega=\omega_1\end{cases}$$

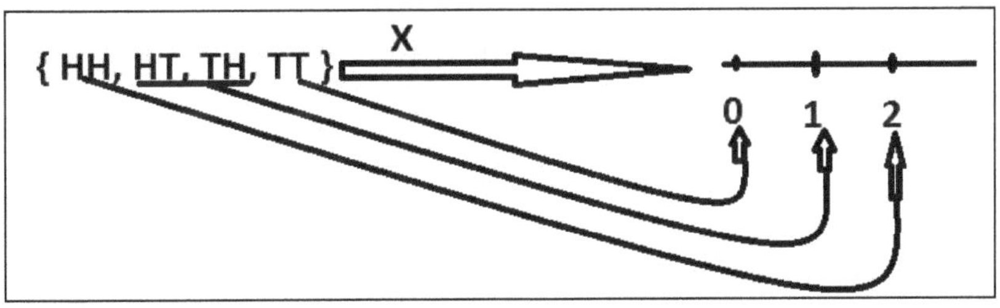

From the above figure we can say that HH corresponds to 2, HT, TH corresponds to 1 and TT corresponds to 0. So here the random variable X is mapping from the sample space to the real line. Here X is taking only three values 0,1 and 2.

A Mathematical Model for Tossing a Coin *N* Times

In order to construct a mathematical model for tossing coin N times, let,

$\Omega=\{\omega=(\omega_1,\ldots\omega_n)|\omega_j=0,1 \text{ for } j = 1,2,\ldots N\}$ $\mathscr{F} = S(\Omega)$. Then Ω has 2^N points and $\mathscr{F}$ has 2^{2^N} sets. For a typical $\omega = (\omega_1,\ldots,\omega_n)$ put,

$$p_\omega = p^{(\omega_1+\omega_2+\ldots+\omega_n)} q^{N-\sum_{i=1}^{N}\omega_i}$$

$$\mathbb{P}[A]=\sum_{\omega\in A} P_\omega, A\in\mathscr{F}=S(\Omega)$$

Now $p_\omega = p^{(\omega_1+\omega_2+\ldots+\omega_n)} q^{N-\sum_{i=1}^{N}\omega_i} \ge 0.$ 0. Obviously $\mathbb{P}[\Omega] = 1$.

$$\mathbb{P}[\Omega]=\sum_{\omega\in\Omega} P_\omega = \sum_{\omega_1}\sum_{\omega_2}\cdots\sum_{\omega_n} p^{\sum_{i=1}^{N}\omega_i} q^{N-\sum_{i=1}^{N}\omega_i}, p+q=1$$

$$=\sum_{\omega_1}\sum_{\omega_2}\cdots\sum_{\omega N-1} p^{\sum_{i=1}^{N-1}\omega_i} q^{N-1-\sum_{i=1}^{N}\omega_i} \sum_{\omega_N=0.1} p^{\omega_n} q^{1-\omega_n}$$

$$=\sum_{\omega_1}\sum_{\omega_2}\cdots\sum_{\omega N-1} p^{\sum_{i=1}^{N-1}\omega_i} q^{N-1-\sum_{i=1}^{N}\omega_i} \ldots=\sum_{\omega_1} p^{\omega_1} q^{1-\omega}=(p+q)=1.$$

Define $X_j(\omega) = \omega_j$, $1 \le j \le N$ for $\omega \in \Omega$. X_j's are called coordinate r.v.s in this model.

Integral

In mathematics, integral is either a numerical value equal to the area under the graph of a function for some interval (definite integral) or a new function the derivative of which is the original function (indefinite integral). These two meanings are related by the fact that a definite integral of any function that can be integrated can be found using the indefinite integral and a corollary to the fundamental theorem of calculus. The definite integral (also called Riemann integral) of a function f(x) is denoted as:

$$\int_a^b f(x)dx.$$

and is equal to the area of the region bounded by the curve (if the function is positive between x = a and x = b) y = f(x), the x-axis, and the lines x = a and x = b. An indefinite integral, sometimes called an anti-derivative, of a function f(x), denoted by:

$$\int f(x)dx.$$

is a function the derivative of which is f(x). Because the derivative of a constant is zero, the indefinite integral is not unique. The process of finding an indefinite integral is called integration.

World of Integrals

Riemann Integration

Let [a, b] be a bounded closed interval on the line $\mathbb{R}$. Let g be a real-valued bounded function on [a, b]. Define a partition of [a, b] by $P := \{x_0, x_1, x_2, \cdots, x_{n-1}, x_n\}$, where $a = x_0 \leq x_1 \leq x_2 \leq \cdots \leq x_{n-1} \leq x_n = b$. Thus, P divides [a, b] into a collection of sub-intervals of the form [xk, xk+1], $k \in \{0, 1, \cdots, n - 1\}$. From these n intervals, arbitrarily select n points $\xi_1, \xi_2, \ldots, \xi_n$, such that $\xi_k \in [x_k-1, x_k]$, and form the sum:

$$S_R := \sum_{k=1}^{n} g(\xi_k)(x_k - x_{k-1}).$$

The limit of the sum as $n \to +\infty$ and $\max_{1\leq k\leq n} |x_k - x_{k-1}| \to 0$, provided it exists, is called the Riemann integral of g on [a, b], and is denoted by:

$$\int_a^b g(x)dx.$$

The Riemann integral is the 'ordinary' integral encountered in everyday mathematics, and can be dealt with (in some situations) by the well-known methods of integration. Sometimes, however, this definition of the integral proves inadequate for applications elsewhere in mathematics. For instance, in probability and statistics, the need for simultaneous treatment of continuous and discrete distributions necessitates generalizing the integral to encompass finite (and possibly, infinite) sums. The Riemann integral cannot achieve this, and this makes for some rather clumsy statements of results. The generalization we look upon, the Riemann-Stieltjes integral, integrates a given function with respect to another. The benefit of the concept is that, among other things, it includes Riemann integrals and finite sums as special cases.

Riemann-Stieltjes Integration

[a, b] is a bounded closed interval on the line $\mathbb{R}$, P is a partition of [a, b] and g: [a, b] → $\mathbb{R}$ is a bounded function. Furthermore, let F be another real-valued function, bounded on [a, b]. For arbitrary points $\xi_1, \xi_2, \ldots \xi_n$, with $\xi_k \in [x_{k-1}, x_k]$, consider the sum:

$$S_{RS} := \sum_{k=1}^{n} g(\xi_k)\left[F(x_k) - F(x_{k-1})\right].$$

As $n \to +\infty$ and $\max_{1 \le k \le n} |x_k - x_{k-1}| \to 0$, the limit of S_{RS}, provided it exists, is called the Riemann-Stieltjes integral of g with respect to F on [a, b], and is denoted by,

$$\int_a^b g(x)\,dF(x).$$

In this case, we say g is (Riemann-Stieltjes) integral, on [a, b], with respect to F, and write g ∈ R ([a, b], F). Before discussing the Riemann-Stieltjes integral in detail, we make some comments on this integral:

- If F is the identity map, i.e., F(x) = x, then the Riemann-Stieltjes integral $\int_a^b g(x)\,dF(x)$ is, simply, the Riemann integral $\int_a^b g(x)\,d(x)$. If we denote the identity map by ψ, i.e., ψ(x) = x, then to say that g is Riemann integrable on [a, b] is the same thing as writing g ∈ R([a, b], ψ).
- If F is non-decreasing and bounded on [a, b], then all functions that are bounded and continuous on [a, b] are integrable with respect to F.
- The Riemann-Stieltjes integral of a bounded function g with respect to F exists even when g has (at most) finitely many discontinuities, provided that F is continuous at all the points of discontinuity of g.
- The converse of item (ii) is true, as well. If g is monotonic on [a, b] and F is continuous on [a, b], then g ∈ R([a, b], F).
- If g ∈ R([a, b], F), and F has a continuous derivative F' on [a, b], then gF' ∈ R([a, b], ψ) and we have:

$$\int_a^b g(x)\,dF(x) = \int_a^b g(x)F'(x)\,dx.$$

Example: Define on [0, 10], the functions g(x) = x and F(x) = 1 − e^{1-x}. Then,

$$\int_0^{10} g(x)\,dF(x) = \int_0^{10} x\,d\left[1 - e^{-x}\right]$$

$$\overset{\text{by}(v)}{=} \int_0^{10} xe^{-x}\,dx$$

$$\overset{\text{by parts}}{=} 1 - 11e^{-10}.$$

The following diagram graphically illustrates the ideas involved in the Riemann integral $\int_0^{10} xe^{-x}dx$ and the Riemann-Stieltjes integral $\int_0^{10} xd\left[1-e^{-x}\right]$.

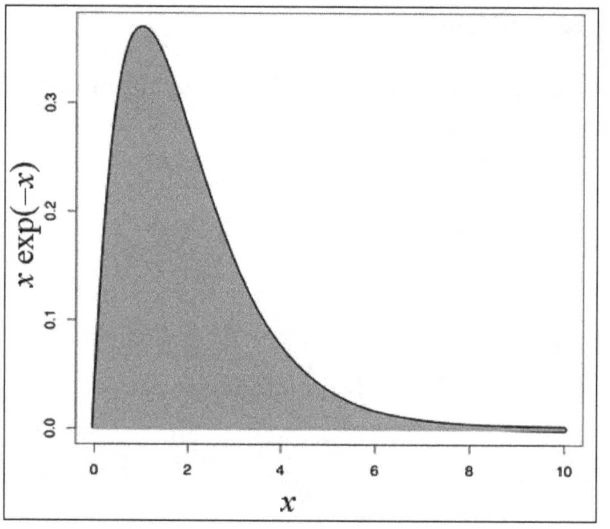

Riemamm integral of [x exp(-x)] with respect to x on [0,10].

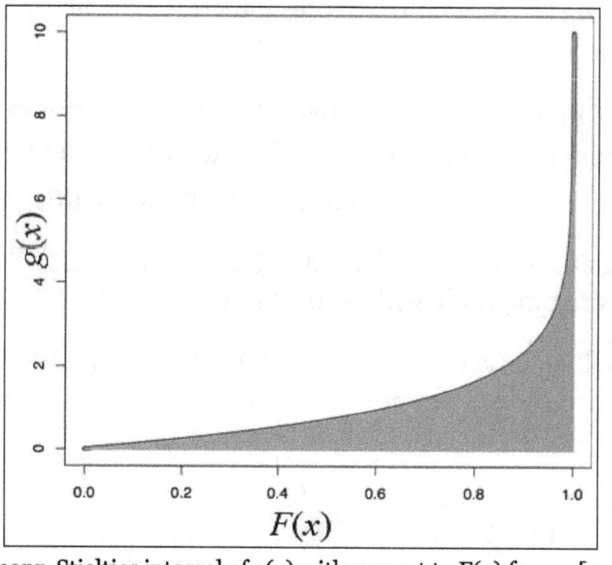

Riemann-Stieltjes integral of g(x) with respect to F(x) for x ϵ [0,10].

We mentioned that one of our primary motivations for studying the Riemann-Stieltjes integral is to 'combine' the otherwise disparate theories of Riemann integration and summation. Above points achieve the connection with Riemann integrals, we, now, see how finite sums are incorporated within the theory:

- Let F is a step function defined on [a, b] with n jumps (discontinuities) of magnitudes p_1,..., p_n, respectively, at the points x_1,..., x_n. If g be such that g and F do not share points of discontinuity, then $g \in R([a, b], F)$ and,

$$\int_a^b g(x)dF(x)=\sum_{j=1}^{n} g(x_j)p_j.$$

Example: On $[-\gamma, \gamma]$ (with $\gamma > 1$), define F by,

$$F(x)=\begin{cases}0, & \text{if } x<0\\ \theta, & \text{if } 0\le x<1\\ 1, & \text{if } x=1,\end{cases}$$

where $\theta \in (0, 1)$, and take g to be the identity map. Clearly, F is a (non-decreasing) step function on [0, 1], with jumps $p_1 = \theta$ and $p_2 = 1 - \theta$ at $x_1 = 0$ and $x_2 = 1$, respectively. Then,

$$\begin{aligned}\int_{-\gamma}^{\gamma} g(x)dF(x) &= \int_{-\gamma}^{\gamma} xdF(x)\\ &\overset{by(vi)}{=} \sum_{j=1,2} x_j p_j\\ &= 1-\theta.\end{aligned}$$

Having seen the great generality of Riemann-Stieltjes integrals, we now consider the details of some of the underlying ideas. We consider some closed and bounded subinterval of $\mathbb{R}$, say [a, b], as the domain of all our functions. Let g and F be bounded functions on [a, b]. We further assume that F is monotonically non-decreasing on [a, b]. $P := \{x_0, x_1, \cdots, x_n\}$ is a partition of [a, b]; $a = x_0 \le x_1 \le \cdots \le x_n = b$. Denote by P[a,b], the collection of all partitions of [a, b]. For a fixed partition P of [a, b], such that $\lim_{n\to\infty} \max_{1\le j\le n} |x_j - x_{j-1}| = 0.$ consider,

$$M_j := \sup\{g(x) : x_{j-1} < x \le x_j\} \text{ and } m_j := \inf\{g(x) : x_{j-1} < x \le x_j\}.$$

Define the upper and lower Darboux sums by,

$$Z_n := \sum_{j=1}^{n} M_j\left[F(x_{j-1})\right] \text{ and } z_n := \sum_{j=1}^{n} m_j\left[F(x_j) - F(x_{j-1})\right].$$

We say that the Riemann-Stieltjes integral of g with respect to F, on [a, b], exists if: $\lim_{n\to\infty} Z_n = \lim_{n\to\infty} z_n$, independently of the choice of the partition. In that case, the common limiting value, is called the Riemann-Stieltjes integral, and is denoted by,

$$\int_a^b g(x)dF(x).$$

Lebesgue-Stieltjes Integration

Let F be a distribution function. Then corresponding to F, we have a unique probability measure, say $\mathbb{P}$, on $B_{\mathbb{R}}$, such that $\mathbb{P}(a, b] = F(b) - F(a)$. Let g be a bounded measurable function on a Borel set S. By a finite partition of S, we mean a collection of disjoint subsets $\{S_i : i = 1,\ldots, n\}$ of S such that $\bigcup_{i=1}^{n} S_i = S$. Also define,

$$Z_n := \sum_{j=1}^{n} M_j \mathbb{P}(S_j) \text{ and } z_n = \sum_{j=1}^{n} m_j \mathbb{P}(S_j),$$

where $M_j := \sup\{g(x) : x \in S_j\}$ and $m_j := \inf\{g(x) : x \in S_j\}$. If $\lim_{n\to\infty} Z_n = \lim_{n\to\infty} z_n$, independently of the choice of the partition, then this common limiting value is called the Lebesgue-Stieltjes integral of g, on S, with respect to $\mathbb{P}$, and is denoted by,

$$\int_s g d\,\mathbb{P} \text{ or } \int_s g(x)\mathbb{P}(dx).$$

If S = [a, b], for some real a, b (a < b), then,

$$\int_{[a,b]} g(x)P(dx) = \int_b^a g(x)dF(x).$$

Of course, F need not be a distribution function, as long as F is non-decreasing and right continuous, in which case F determines (uniquely) a measure μ, satisfying μ (a, b] = F(b)−F(a). The Lebesgue-Stieltjes integral, of a measurable function g, with respect to μ can be defined analogously, and will equal the Riemann-Stieltjes integral (if it is defined) with respect to F. Let us take a look at some of the ramifications of the definition of Lebesgue-Stieltjes integral:

- The particular choice F(x) = x yields the Lebesgue measure $\mathbb{L}$ (say), and then the Lebesgue-Stieltjes integral is the Lebesgue integral on the measure space $(\mathbb{R}, B_{\mathbb{R}}, \mathbb{L})$.
- Let X be a (bounded) random variable on some probability space $(\Omega, A, \mathbb{P})$, taking values in the interval [a, b]. Let F_X be the distribution function of X, and this corresponds to a unique probability measure $\mathbb{P}_X$ – the law or distribution of X. The expectation of X, provided it exists, is defined by,

  $$\mathbb{E}[X] := \int_\Omega X(\omega)\mathbb{P}(d\omega).$$

  This Lebesgue integral also equals the following Lebesgue-Stieltjes and RiemannStieltjes integrals:

  $$\mathbb{E}[X] = \int_\Omega X(\omega)\mathbb{P}(d\omega) = \int_{[a,b]} x\mathbb{P}(dx) = \int_b^a x dF_x(x).$$

- If F is a step function with jumps p_1, p_2 at the points $x_1, x_2, \ldots$, then, the Riemann-Stieltjes integral reduces to a sum. Hence, the corresponding Lebesgue-Stieltjes integral, too, reduces to a sum.
- The true power of Riemann-Stieltjes integration is revealed when both Riemann integration and summation prove inadequate.

Example: Let X be a random variable, uniformly distributed over the Cantor set C. If we write the probability space over which X is defined as $(\Omega, \mathbb{A}, \mathbb{P})$, then $\mathbb{P}\{X \in [0, 1]\} = 1$, since $C \subset [0, 1]$ and $\mathbb{P}\{X \in C\} = 1$. So, the expectation of X can be evaluated as,

$$\mathbb{E}[X] = \int_\Omega X(\omega)\mathbb{P}(d\omega) = \int_{[0,1]} x\mathbb{U}_C(dx) = \int_0^1 x dF_x(x).$$

where $F_X(x) := \mathbb{U}_C(-\infty, x]$ is the distribution function corresponding to $\mathbb{U}_C$. Note that F_X is the Cantor function, which is continuous but not absolutely continuous (with respect to Lebesgue measure). This implies that neither the PDF nor the PMF of X exists, and so the last Riemann-Stieltjes integral cannot be simplified to a Riemann integral or a sum. We can, however, calculate E[X]. The integration by parts formula implies:

$$\int_0^1 x dF_X(x) = 1 \cdot F_X(1) - 0 \cdot F_X(0) - \int_0^1 F_X(x) dx = 1 - \int_0^1 F_X(x) dx.$$

The last integral is an ordinary Riemann integral, and it exists as F_X is continuous. Remember that the Riemann integral of a function is the 'area under the curve.' As the Cantor function divides the 1 × 1 square in the positive quadrant of the x-F(x) plane into two equal halves, we have

$\int_0^1 F_X(x) dx = \frac{1}{2} = \frac{1}{2}$, and,

$$\mathbb{E}[X] = 1 - \frac{1}{2} = \frac{1}{2}.$$

Extension of the Riemann-Stieltjes Integration

The Riemann-Stieltjes integral $\int_a^b g(x) dF(x)$ can be extended to accommodate the case(s) when one or both of the limits of integration tend to one of the infinities. We define,

$$\int_{-\infty}^{+\infty} g(x) dF(x) := \lim_{\substack{a \to -\infty \\ b \to +\infty}} \int_a^b g(x) dF(x),$$

provided $\int_a^b g(x) dF(x)$ exists.

Previously, we defined the Riemann-Stieltjes integral, with respect to a function F. Assume that F is a distribution function. Let g be a function continuous on [a, b]. Consider a partition P of [a, b], given by P := $\{x_0, x_1, \cdots, x_n\}$, for a = $x_0 \le x_1 \le \cdots \le x_n$ = b. Define the sum:

$$S_{RS} := \sum_{j=1}^{n} g(\xi_j)\left[F(x_j) - F(x_{j-1})\right],$$

where $\xi_j \in [x_{j-1}, x_j]$. The Riemann-Stieltjes integral of g, on [a, b], with respect to F is defined to be the limit of SRS, as n →∞ and $\max_{1\le j\le n} |x_j - x_{j-1}| \to 0$, provided the limit exists, and is denoted by,

$$\int_a^b g(x) dF(x).$$

In this case, we say that g is integrable with respect to F on [a, b], and write g ∈ R([a, b], F).

This definition can be extended to infinite intervals. If g is integrable, over [a, b], with respect to F then the integral $\int_{-\infty}^{\infty} g(x) dF(x)$ is defined to be the Riemann-Stieltjes integral $\int_a^b g(x) dF(x)$ a →-∞ and b →∞, provided $\lim_{n\to\infty} \max_{1\le j\le n} |x_j - x_{j-1}| = 0$; we write g ∈ R(ℝ, F).

We, now, look at a few properties of the Riemann-Stieltjes integral:

(i) $\int_a^b [g_1(x) + g_2(x)] dF(x) = \int_a^b g_1(x) dF(x) + \int_a^b g_2(x) dF(x).$

(ii) $\int_a^b g(x)d[F_1(x)+F_2(x)]=\int_a^b g(x)dF_1(x)+\int_a^b g(x)dF_2(x).$

(iii) $\int_a^b g(x)dF(x)=\int_a^c g(x)dF(x)+\int_c^b g(x)dF(x)$, where, $a<c<b$.

(iv) If $g \in R([a, b], F)$ and $m \le g \le M$ (for constants m and M), and φ is a continuous function defined on $[m, M]$, then $h := \varphi \circ g \in R([a, b], F)$. These results can be easily proved from the definition.

Theorem: (Helly's Theorem) Consider the bounded interval [a, b], and let g be a function bounded and continuous on [a, b]. Let $\{F_n\}_{n\in\mathbb{Z}+}$ be a sequence of bounded and non-decreasing functions on [a, b], converging to another (bounded) function F, point-wise at every point of continuity of F. Then,

$$\lim_{n\to\infty}\int_a^b g(x)dF_n(x)=\int_a^b g(x)dF(x).$$

Before we prove Helly's Theorem, we consider an application.

Example: On [0, 1], define $F_n(x) := x^{3+\frac{2}{\sqrt{n}}}\ (\forall_n \in \mathbb{N})$, $F(x) := x^3$ and $g(x) := x^2$. As $n \to \infty$, $F_n \to F$, point-wise on [0, 1]. Now,

$$\int_0^1 x^2 dF_n(x)=\int_0^1 x^2 d\left[x^{3+\frac{2}{\sqrt{n}}}\right]=\frac{3+\frac{2}{\sqrt{n}}}{5+\frac{2}{\sqrt{n}}}\to\frac{3}{5}, \text{as } n\to\infty,$$

and,

$$\int_0^1 x^2 dF(x)=\int_0^1 x^2 d\left[x^3\right]=\frac{3}{5}.$$

Thus, $\lim_{n\to\infty}\int_0^1 x^2 dF_n(x)=\int_0^1 x^2 dF(x)$, as asserted by Helly's Theorem.

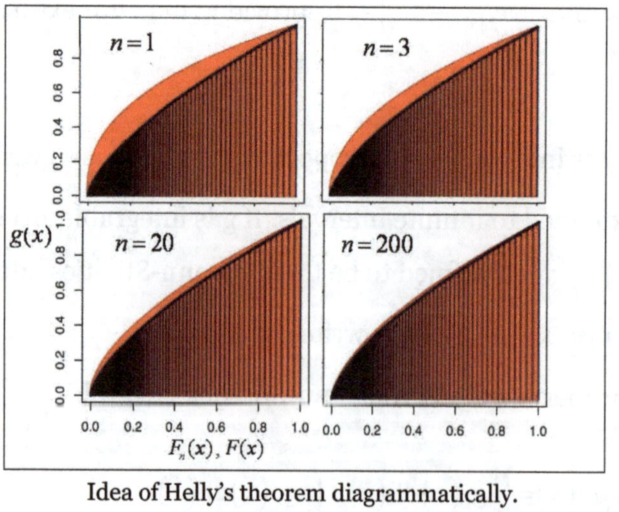

Idea of Helly's theorem diagrammatically.

Figure shows why this must be the case. We calculate the integral $\int_0^1 x^2 dF_n(x)$ for n = 1, 3, 20, and 200. The red shaded area is the limiting value of the integral i.e. area under the integral $\int_0^1 x^2 d(x^3)$. Now it is clear that as n increases, the black shaded region converges to the red region giving a diagrammatic view of Helly's theorem. Note that the functions used in this example satisfy the conditions stated in Helly's theorem.

Proof: (of Helly's theorem) As all the F_n's are non-decreasing on [a, b], so is F. Let $x_0, x_1, \ldots, x_k$ ($a = x_0 < x_1 < \ldots < x_k = b$) be the points of continuity of F. Then $\forall \varepsilon > 0$,

$$\begin{aligned}\int_a^b g(x)dF(x) &= \sum_{i=0}^{k-1}\int_{x_i}^{x_{i+1}} g(x)dF(x)\\ &= \sum_{i=0}^{k-1}\int_{x_i}^{x_{i+1}}\left[g(x)-g(x_i)\right]dF(x)+\sum_{i=0}^{k-1}\int_{x_i}^{x_{i+1}} g(x_i)\left[F(x_{i+1})-F(x_i)\right]\\ &< \frac{\theta_{1\varepsilon}}{3}+\sum_{i=0}^{k-1} g(x_i)\left[F(x_{i+1})-F(x_i)\right].\end{aligned}$$

Similarly,

$$\int_a^b g(x)dF_n(x) < \frac{\theta_{2\varepsilon}}{3}+\sum_{i=0}^{k-1} g(x_i)\left[F_n(x_{i+1})-F_n(x_i)\right].$$

Since, $F_n(x) \to F(x)$, point-wise at continuity points of F, the absolute difference between the integrals in above equations is less than ε, for large enough n. Thus, as $n \to \infty$, the proof follows. It is possible to extend Helly's Theorem to the case when the interval of integration is ℝ.

Theorem: (Helly-Bray Theorem) If (a) g is a continuous function, (b) $F_n(x) \to F(x)$, as $n \to \infty$, point-wise at every point of continuity of F, and (c) $\forall \varepsilon\ (> 0), \exists A (> 0)$ such that,

$$\int_{-\infty}^{-A} |g(x)| dF(x) + \int_A^{\infty} |g(x)| dF(x) < \varepsilon,\ \forall_n = 1, 2\ldots,$$

$$\lim_{n\to\infty} \int_{-\infty}^{\infty} g(x)dF_n(x) = \int_{-\infty}^{\infty} g(x)dF_n(x).$$

Here, we consider some important theorems on Riemann-Stieltjes integrals, which have direct applications to exact computation of integrals. We do not prove the theorems, but illustrate through examples.

Theorem: (Dominated Convergence Theorem) If $\{g_n\}_{n\in Z+}$ is a sequence of real-valued functions on [a, b], converging (point-wise) to another function g, and for all $n \in \mathbb{Z}^+$,

$$|g_n(x)| \le G(x),\ \forall x \in [a, b],$$

where $G \in R([a, b], F)$, then all of $g, g_1, g_2, \ldots \in R([a, b], F)$ and

$$\lim_{n\to\infty}\int_a^b g_n(x)dF(x) = \int_a^b g(x)dF(x).$$

Example: On [0, 2], define $g_n(x) := e^{-\left(x+\frac{1}{\sqrt{n}}\right)} (\forall_n \in \mathbb{N})$, $G(x) := e^{-x}$ and $F(x) := \frac{1-e^{-x}}{1-e^{-2}}$. As $g_n \to G$, point-wise on [0, 2], set g ≡ G. [Obviously, for all $n \in \mathbb{N}$, we have $|g_n(x)| \le G(x)$, $\forall x \in [a, b]$.] The Dominated Convergence Theorem is easily verified in this case:

$$\lim_{n\to\infty}\int_0^2 g_n(x)dF(x) = \lim_{n\to\infty}\int_0^2 e^{-\left(x+\frac{1}{\sqrt{n}}\right)} d\left[\frac{1-e^{-x}}{1-e^{-2}}\right]$$

$$= \lim_{n\to\infty}\int_0^2 \frac{e^{-\left(x+\frac{1}{\sqrt{n}}\right)}e^{-x}}{1-e^{-2}}dx$$

$$= \lim_{n\to\infty}\frac{1+e^{-2}}{2e^{1/\sqrt{n}}} = \frac{1+e^{-2}}{2}.$$

The limit and the integral are equal, as asserted by the Dominated Convergence Theorem.

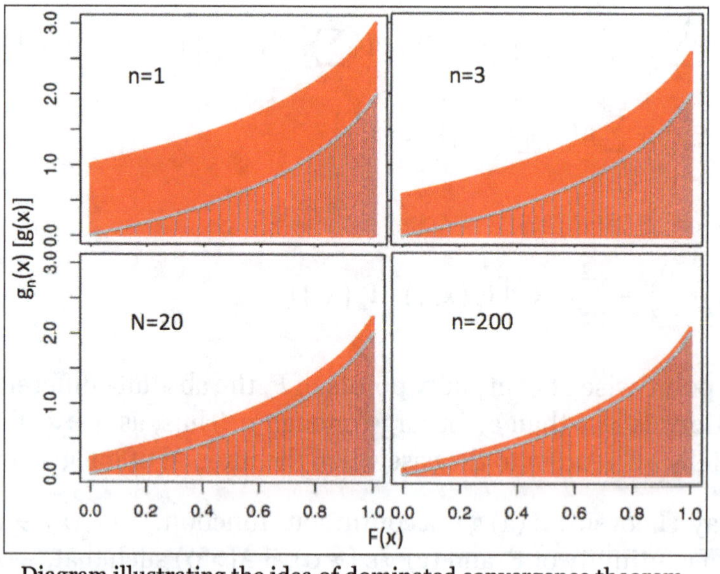

Diagram illustrating the idea of dominated convergence theorem.

Theorem: (Differentiation under the Integral Sign) If for continuity points x of a function g(·, t), for a fixed value of t, (i) the partial derivative $\frac{\partial g(x,t)}{\partial t}$ exists, and (ii) $\left|\frac{\partial g(x,t)}{\partial t}\right| \le G(x)$, $\forall x$ [where $G \in R(\mathbb{R}, F)$], then,

$$\frac{\partial}{\partial t}\int_{-\infty}^{\infty} g(x,t)dF(x) = \int_{-\infty}^{\infty}\frac{\partial g(x,t)}{\partial t}dF(x).$$

Example: Over [0,∞), define $g(x, t) := e^{-tx}$ and $F(x) := 1 - e^{-x}$. Then,

$$\frac{\partial}{\partial t}\int_0^{\infty} g(x,t)dF(x) = \frac{\partial}{\partial t}\int_0^{\infty} e^{-tx}d\left[1-e^{-x}\right] = \frac{\partial}{\partial t}\left[\frac{1}{1+t}\right] = \frac{1}{(1+t)^2},$$

$$\int_0^{\infty}\frac{\partial g(x,t)}{\partial t}dF(x) = \int_0^{\infty}(-x)e^{-tx} = -\int_0^{\infty} xe^{-(1+t)x}dx = -\frac{1}{(1+t)^2}.$$

Theorem: (Fubini's Theorem) If (i) g(x, ·) is continuous (in the second argument) on (a, b), for all $x \in \mathbb{R}$, (ii) $|g(x, t)| \le G_1(x)$, for all $t \in (a, b)$ and all $x \in R$, and (iii) $\int_{-\infty}^{\infty} |g(x, t)|dt \le G_2(x)$, for all $x \in \mathbb{R}$, where $G_1, G_2 \in R(\mathbb{R}, F)$, then,

$$\int_{-\infty}^{\infty}\left[\int_{-\infty}^{\infty}|g(x,t)|dF(x)\right]dt=\int_{-\infty}^{\infty}\left[|g(x,t)|dt\right]dF(x).$$

Example: For some fixed real number θ, define on $[0,\infty)$, the functions $g(x, t) := \theta e^{-(\theta t+x)}$, $G_1(x) = G_2(x) := e -x$ and $F(x) := x$. Then,

$$\int_0^{\infty}\left[\int_0^{t}g(x,t)dF(x)\right]dt=\int_0^{\infty}\left[\int_0^{t}\theta e^{-(\theta t+x)}dx\right]dt$$
$$=\int_0^{\infty}\theta e^{-\theta t}\left(1-e^{-t}\right)dt$$
$$=\frac{1}{1+\theta}.$$

and
$$\int_0^{\infty}\left[\int_x^{\infty}g(x,t)dt\right]dF(x)=\int_0^{\infty}\left[\int_x^{\infty}\theta e^{-(\theta t+x)}dt\right]dx$$
$$=\int_0^{\infty}e^{-x(1+\theta)}dx$$
$$=\frac{1}{1+\theta}.\blacklozenge$$

Characteristic Function

The distribution or law of a real-valued random variable X, defined on some probability space $(\Omega, \mathcal{A}, \mathbb{P})$, is a probability measure L_X on $\mathscr{B}_{\mathbb{R}}$, such that for any (linear) Borel set $\mathbb{B}$ (i.e., for any $\mathbb{B} \in \mathscr{B}_{\mathbb{R}}$),

$$\mathbb{L}_X(\mathbb{B}):=\mathbb{P}\left(X^{-1}(\mathbb{B})\right)\overset{\text{def}}{=}\mathbb{P}\{X\in\mathbb{B}\}.$$

L_X is a probability measure defined on $\mathscr{B}_{\mathbb{R}}$, and hence can be uniquely represented by a non-decreasing, right continuous function F_X given by,

$$F_X(x) : \mathbb{L}_X(\infty, x] = \mathbb{P}\{X \le x\}, \forall x \in \mathbb{R}.$$

Furthermore, this distribution function characterizes the distribution $\mathbb{L}_X$ of X: F_X corresponds to a single probability measure on $\mathscr{B}_{\mathbb{R}}$, and that is $\mathbb{L}_X$. We can define the moments of a distribution as:

$$\mu_r:=\mathbb{E}\left[X^r\right]=\int_{\mathbb{R}}x^r\mathbb{L}_x(dx)=\int_{-\infty}^{\infty}x^r dF_X(X).$$

But the sequence of moments $\{\mu_r\}_{r\in N}$, even when they all exist, do not necessarily characterize a distribution. The moment generating function (MGF) of a distribution $\mathbb{L}_X$, defined as a function on some interval containing 0, is given by,

$$M_X(t):=\mathbb{E}\left[\exp(tX)\right]=\int_{\mathbb{R}}\exp(tx)\mathbb{L}_X(dx)=\int_{-\infty}^{\infty}\exp(tx)dF_{X(x)}.$$

It is easy to see that the MGF of a distribution can be expressed as,

$$M_X(t)=1+t\cdot\frac{\mathbb{E}[X]}{1!}+t^2\cdot\frac{\mathbb{E}\left[X^2\right]}{2!}+t^3\cdot\frac{\mathbb{E}\left[X^3\right]}{3!}+...,$$

and its existence depends on the convergence of this series. Is it possible to characterize a distribution uniquely, such that the characterization works for all distributions? The answer is "yes".

The characteristic function of a random variable X, with distribution L_X (or equivalently, distribution function F_X) is a function $\varphi_X : \mathbb{R} \to \mathbb{C}$ defined by,

$$\phi_x(t) := \mathbb{E}[\exp(itX)] = \int_{\mathbb{R}} \exp(itx)\mathbb{L}_X(dx) = \int_{-\infty}^{\infty} \exp(itx)dF_{X(x)},$$

where $i^2 = -1$.

The Euler identity $\exp(i\theta) = \cos\theta + I\sin\theta$, gives simple expressions for the real and imaginary parts of the characteristic function:

$$\Re(\phi_x(t)) = \mathbb{E}[\cos(tX)] \text{ and } \Im(\phi_X(t)) = \mathbb{E}[\sin(tX)].$$

The next few theorems list some basic properties of characteristic functions which make them suitable for applications.

Theorem: Let X be a random variable with distribution function F and ChF φ. Then,

(i) $\varphi(0) = 1$.

(ii) $|\varphi(t)| \leq 1, \forall t \in \mathbb{R}$.

(iii) $\varphi(.)$ is a Hermitian function: $\varphi(-t) = \overline{\phi(t)}$.

Proof: (i) $\varphi(0) = E[\exp(0)] = 1$.

(ii) $|\phi(t+h) - \phi(t)| = \left|\int_{-\infty}^{\infty} \exp(itX)\right| = \int_{-\infty}^{\infty} |\exp(itx)| dF(x) = \int_{-\infty}^{\infty} dF(x) = 1.$

(iii) Using the relations $\cos\theta = \cos(-\theta)$ and $\sin(-\theta) = -\sin\theta$, and the definition of complex conjugation, we have,

$$\overline{\phi(t)} \overset{\text{def}}{=} \mathbb{E}[\cos(tX)] - i\mathbb{E}[\sin(tX)] = \mathbb{E}[\cos(-tX)] + i\mathbb{E}[\sin(-tX)] = \phi(-t).$$

Theorem: Every ChF φ is uniformly continuous on $\mathbb{R}$.

Proof: For any $t \in \mathbb{R}$,

$$\begin{aligned} |\phi(t+h) - \phi(t)| &= \left|\int_{-\infty}^{\infty} [\exp(i(t+h)x) - \exp(itx)] dF(x)\right| \\ &\leq \int_{-\infty}^{\infty} |\exp(i(t+h)x) - \exp(itx)| dF(x) \\ &= \int_{-\infty}^{\infty} \underbrace{|\exp(itx)|}_{-1} |\exp(ihx) - 1| dF(x) \\ &= \int_{-\infty}^{\infty} |\exp(ihx) - 1| dF(x). \end{aligned}$$

Now, $|\exp(ihx) - 1| \leq 2$, and $\lim_{h\to 0} |\exp(ihx) - 1| \to 0$. So, a simple application of the dominated convergence theorem shows,

$$\int_{-\infty}^{\infty}\left|\exp(ihx)-1\right|dF(x)\to 0,$$

which proves uniform continuity φ.

Theorem If φ_1, φ_2,..., φ_n are characteristic functions, corresponding to distribution functions F_1, F_2,..., F_n, respectively, and if a_1, a_2,..., a_n are non-negative real numbers such that $\sum_{j\ 1}$ 1, then the function φ, define,

$$\phi(t)=\sum_{j=1}^{n}a_j\phi_j(t),$$

is also a characteristic function.

Proof: Define $F(x):=\sum_{j=1}^{n}a_jF_j(x)$, for each real x. Then F defines a distribution function as well. The characteristic function of F is,

$$\int_{-\infty}^{\infty}\exp(itx)dF(x)=\int_{-\infty}^{\infty}\exp(itx)d\left[\sum_{j=1}^{n}a_jF_j(x)\right]$$

$$=\sum_{j=1}^{n}\left[\int_{-\infty}^{\infty}\exp(itx)dF_j(x)\right]$$

$$=\sum_{j=1}^{n}a_j\phi_j(t)=\phi(t).$$

Hence, φ is the characteristic function of F.

Corollary: If φ is a characteristic function, then so is $\overline{\phi}$. Also, the real part of φ, given by,

$$\text{Re}(\phi(t))=\frac{1}{2}\cdot\phi(t)+\frac{1}{2}\cdot\overline{\phi(t)},$$

is a characteristic function, by above Theorem.

If F is a distribution function, and a and b are real constants, with b nonzero, then it is easy to check that G(x) := $F\left(\frac{x-a}{b}\right)$, is also a distribution function. Furthermore, if φ is the ChF of F, then γ(t) := exp(it a)φ(bt) is the ChF of G. For our next result, we recall the idea of symmetrically distributed random variables.

Definition: Let X be a random variable with distribution function F. The function $\tilde{F}$, defined by $\tilde{F}$ (x) := 1 – F(-x⁻), is also a distribution function, called the conjugate of F. Moreover, $\tilde{F}$ is the distribution function of-X.

Definition: Let X be a random variable with distribution function F. If X and -X have the same distribution (function), i.e., F = $\tilde{F}$, then X is said to be distributed symmetrically about the origin.

Theorem: A distribution function is symmetric if and only if its ChF is real and even.

Proof: If φ is real, then $\phi(t)=\overline{\phi(t)}$, for all t. By Theorem $\varphi(t)=\overline{\phi(t)}$, and so φ is even, whenever it is real. Also, by the properties of the trigonometric functions, we can easily deduce that φ is the ChF of -X. So, the characteristic functions of X and -X are the same, and by the uniqueness of characteristic functions, it follows that the distributions of X and X are the same. Thus, by Definition above the 'if' part of the theorem is proved. For the 'only if' part, observe that all the steps are reversible.

Theorem: If X and Y are symmetrically distributed random variables, with distribution functions F and G, respectively, then X Y is also symmetrically distributed.

Proof: Let the characteristic function of F be φ and that of G be γ. As X and Y are independent, the distribution function of Z := X + Y is given by the convolution.

$$F * G(z) := \int_{-\infty}^{\infty} F(z-y)\,dG(y) = \int_{-\infty}^{\infty} G(z-x)\,dF(x).$$

The characteristic function η of Z can easily be seen to be the product of the characteristic functions of X and Y:

$$\begin{aligned}\eta(t) &\overset{\text{def}}{=} \mathbb{E}[\exp(itZ)] \\ &= \mathbb{E}[\exp(it(X+Y))] \\ &= \mathbb{E}[\exp(itX)]\mathbb{E}[\exp(itY)][\because X \perp Y] \\ &\overset{\text{def}}{=} \phi(t)\gamma(t).\end{aligned}$$

As X and Y are symmetrically distributed, φ and γ are real-valued functions, and η, the product of two real-valued functions, is also real-valued. So, by Theorem, Z is symmetric.

Probability Distribution

A probability distribution is a statistical function that describes all the possible values and likelihoods that a random variable can take within a given range. This range will be bounded between the minimum and maximum possible values, but precisely where the possible value is likely to be plotted on the probability distribution depends on a number of factors. These factors include the distribution's mean (average), standard deviation, skewness, and kurtosis.

How Probability Distributions Work

Perhaps the most common probability distribution is the normal distribution, or "bell curve," although several distributions exist that are commonly used. Typically, the data generating process of some phenomenon will dictate its probability distribution. This process is called the probability density function. Probability distributions can also be used to create cumulative distribution functions (CDFs), which adds up the probability of occurrences cumulatively and will always start at zero and end at 100%.

Academics, financial analysts and fund managers alike may determine a particular stock's probability distribution to evaluate the possible expected returns that the stock may yield in the future. The stock's history of returns, which can be measured from any time interval, will likely be composed of only a fraction of the stock's returns, which will subject the analysis to sampling error. By increasing the sample size, this error can be dramatically reduced.

Types of Probability Distributions

There are many different classifications of probability distributions. Some of them include the normal distribution, chi square distribution, binomial distribution, and Poisson distribution. The different probability distributions serve different purposes and represent different data generation processes. The binomial distribution, for example, evaluates the probability of an event occurring several times over a given number of trials and given the event's probability in each trial. and may be generated by keeping track of how many free throws a basketball player makes in a game, where 1 = a basket and 0 = a miss. Another typical example would be to use a fair coin and figuring the probability of that coin coming up heads in 10 straight flips. A binomial distribution is discrete, as opposed to continuous, since only 1 or 0 is a valid response.

The most commonly used distribution is the normal distribution, which is used frequently in finance, investing, science, and engineering. The normal distribution is fully characterized by its mean and standard deviation, meaning the distribution is not skewed and does exhibit kurtosis. This makes the distribution symmetric and it is depicted as a bell-shaped curve when plotted. A normal distribution is defined by a mean (average) of zero and a standard deviation of 1.0, with a skew of zero and kurtosis = 3. In a normal distribution, approximately 68% of the data collected will fall within +/- one standard deviation of the mean; approximately 95% within +/- two standard deviations; and 99.7% within three standard deviations. Unlike the binomial distribution, the normal distribution is continuous, meaning that all possible values are represented (as opposed to just 0 and 1 with nothing in between).

Probability Distributions used in Investing

Stock returns are often assumed to be normally distributed but in reality, they exhibit kurtosis with large negative and positive returns seeming to occur more than would be predicted by a normal distribution. In fact, because stock prices are bounded by zero but offer a potential unlimited upside, the distribution of stock returns has been described as log-normal. This shows up on a plot of stock returns with the tails of the distribution having greater thickness.

Probability distributions are often used in risk management as well to evaluate the probability and amount of losses that an investment portfolio would incur based on a distribution of historical returns. One popular risk management metric used in investing is value-at-risk (VaR). VaR yields the minimum loss that can occur given a probability and time frame for a portfolio. Alternatively, an investor can get a probability of loss for an amount of loss and time frame using VaR. Misuse and over-reliance on VaR has been implicated as one of the major causes of the 2008 financial crisis.

Example of a Probability Distribution

As a simple example of a probability distribution, let us look at the number observed when rolling two standard six-sided dice. Each die has a 1/6 probability of rolling any single number, one through six, but the sum of two dice will form the probability distribution depicted in the image below. Seven is the most common outcome (1+6, 6+1, 5+2, 2+5, 3+4, 4+3). Two and twelve, on the other hand, are far less likely (1+1 and 6+6).

3

Statistical Inference and Estimation

The process of using data analysis to conclude the parameters of populations or scientific truths on the basis of random sampling is known as statistical inference. This chapter elaborates nonparametric inference, bayesian inference, point estimation, criteria of estimation, unbiased estimation, etc. All the diverse principles of statistical inference and estimation have been carefully analyzed in this chapter.

Statistical Inference

Statistical inference is the process of analysing the result and making conclusions from data subject to random variation. It is also called inferential statistics. Hypothesis testing and confidence intervals are the applications of the statistical inference. Statistical inference is a method of making decisions about the parameters of a population, based on random sampling. It helps to assess the relationship between the dependent and independent variables. The purpose of statistical inference to estimate the uncertainty or sample to sample variation. It allows us to provide a probable range of values for the true values of something in the population. The components used for making statistical inference are:

- Sample Size.
- Variability in the sample.
- Size of the observed differences.

Types of Statistical Inference

There are different types of statistical inferences that are extensively used for making conclusions. They are:

- One sample hypothesis testing.
- Confidence Interval.
- Pearson Correlation.
- Bi-variate regression.
- Multi-variate regression.
- Chi-square statistics and contingency table.
- ANOVA or T-test.

Statistical Inference Procedure

The procedure involved in inferential statistics are:

- Begin with a theory.
- Create a research hypothesis.
- Operationalize the variables.
- Recognize the population to which the study results should apply.
- Formulate a null hypothesis for this population.
- Accumulate a sample from the population and continue the study.
- Conduct statistical tests to see if the collected sample properties are adequately different from what would be expected under the null hypothesis to be able to reject the null hypothesis.

Statistical Inference Solution

Statistical inference solutions produce efficient use of statistical data relating to groups of individuals or trials. It deals with all characters, including the collection, investigation and analysis of data and organizing the collected data. By statistical inference solution, people can acquire knowledge after starting their work in diverse fields. Some statistical inference solution facts are:

- It is a common way to assume that the observed sample is of independent observations from a population type like Poisson or normal.
- Statistical inference solution is used to evaluate the parameter(s) of the expected model like normal mean or binomial proportion.

Importance of Statistical Inference

Inferential Statistics is important to examine the data properly. To make an accurate conclusion, proper data analysis is important to interpret the research results. It is majorly used in the future prediction for various observations in different fields. It helps us to make inference about the data. The statistical inference has a wide range of application in different fields, such as:

- Business Analysis,
- Artificial Intelligence,
- Financial Analysis,
- Fraud Detection,
- Machine Learning,
- Share Market,
- Pharmaceutical Sector.

Statistical Inference Examples

An example of statistical inference is given below. From the shuffled pack of cards, a card is drawn. This trial is repeated for 400 times, and the suits are given below:

Suit	Spade	Clubs	Hearts	Diamonds
No. of times drawn	90	100	120	90

While a card is tried at random, then what is the probability of getting a:

- Diamond cards.
- Black cards.
- Except for spade.

By statistical inference solution,

Total number of events = 400

i.e.,90+100+120+90=400

- The probability of getting diamond cards:

  Number of trials in which diamond card is drawn = 90,

  Therefore, P(diamond card) = 90/400 = 0.225.

- The probability of getting black cards:

  Number of trials in which black card showed up = 90+100 =190,

  Therefore, P (black card) = 190/400 = 0.475.

- Except for spade:

  Number of trials other than spade showed up = 90+100+120 =310,

  Therefore, P (except spade) = 310/400 = 0.775.

Nonparametric Inference

Nonparametric statistics is a method that makes statistical inference without regard to any underlying distribution. The method fits a normal distribution under no assumptions. Habitually, the approach uses data that is often ordinal because it relies on rankings rather than on numbers. Nonparametric statistics can be contrasted with parametric statistics. The latter approach makes explicit assumptions about the distribution of observed data and estimates the parameters of the distribution using the same data.

Parametric Procedures: The Traditional Practice with Limitations

Most of the traditional statistical tools are based on the parametric assumption: the data at hand can be thought of as generated by some well-known distribution like normal, exponential or Poisson. The parameter(s) of the distributions are assumed unknown and in a parametric inference problem we try to put some idea about the unknown parameter in the form of estimation or testing or confidence interval estimation. However, most of the times the normal distribution is used as the underlying population.

The assumption of normality is often justified by the Central Limit Theorem, which ensures closeness to normality of certain statistics for large enough sample sizes. Other distributions are also important in different fields of applications. For example lifetimes of physical systems are often characterized by exponential, Weibull or Gamma distributions. In such cases the interest lies in knowing the expected failure times and the mentioned distributions are used to characterize the lifetime distribution. The lifetime distribution is also of interest in the field of certain medical trials, where the goal is to give an idea about the length of the life after certain treatments. Exponential, lognormal or Weibull are often used to model the underlying lifetime distribution so as to enable appropriate inferential procedures. Again, in the analysis of economic data, Pareto or lognormal are the appropriate to capture the features of the income distribution.

However, in practice, most of the experiments are of complex nature, affected by a number of factors and hence the generated data might not be identified by a well-known distribution. Although for large data sets, normality assumption can be made for analysis but often the amount of data is small and hence requires assumption of an appropriate distribution for analysis. In fact, there are no exploratory methods to find the appropriate underlying distribution or to differentiate among the available choices. Therefore, a traditional statistician has to assume a distribution for the data which captures only certain features of the data to find without caring for the quality of the inference.

Nonparametric and Distribution Free Procedures

Although the terms nonparametric and distribution-free are invariably used, but these actually indicate statistical procedures, which are used in the absence of any assumption of the underlying distribution. In the words of Bradley:

> "The terms nonparametric and distribution-free are not synonymous, Popular usage, however, has equated the terms, Roughly speaking, a nonparametric test is one which makes no hypothesis about the value of a parameter in a statistical density function, whereas a distribution-free test is one which makes no assumptions about the precise form of the sampled population".

We start with the definition of these two important concepts and identify the differences subsequently.

Distribution Free Statistic

Let X_i, i = 1, 2,,,, n be n variables with the joint distribution F, where $F \in \mathcal{F}$, a class of joint distributions. Then a statistic $T = T(X_1,.., X_n)$ is said to be distribution free over $\mathcal{F}$, if the distribution of T remains the same for every possible $F \in \mathcal{F}$.

For example, if F is the joint distribution of n iid normal variables with unknown mean μ and variance unity. Then the statistic $Z=\sum_{i=1}^{n}\left(X_i-\bar{X}\right)^2$ has a χ^2 distribution with n-1 degrees of freedom and hence is distribution free over $\mathcal{F}$.

However, in the above example, the distribution free statistic is actually ancillary in parametric terminology, that is, they are independent of the indexing parameter of the joint distribution. But they are not nonparametric statistics, since their distributions vary for different joint distributions. For instance, the distribution of Z above is no longer χ^2 when the joint distribution is different from normal.

Nonparametric Distribution Free Statistic

A statistic $T = T(X_1,.., X_n)$ is nonparametric distribution free over $\mathcal{F}$, if the distribution of T does not depend on any $F \in \mathcal{F}$.

For example, if F is the joint distribution of n iid continuous random variables having symmetry at the origin, then the statistic $Z=\sum_{i=1}^{n} I(X_i > 0)$, has a binomial distribution with parameters n and success probability $\theta_F = P_F(X_1 > 0)$. Now due to symmetry at the origin, $\theta_F = \frac{1}{2}$ and hence the distribution of Z remains the same (i.e. Binomial $(n, \frac{1}{2})$) whatever the distribution F be. Thus Z is nonparametric distribution free. However, for the sake of simplicity, we use the term nonparametric to refer statistical procedures based on a nonparametric distribution free statistic.

Advantages, Disadvantages and Recommendation

The main problem with parametric inference is that if the assumed distribution is not correct, then all the efforts might went into vein and the best (i.e. efficient) can become the "worst". Therefore, nonparametric procedures enjoy the following advantages:

- Nonparametric procedures are based on fewer assumptions about the underlying populations of the data. Therefore, these procedures can accommodate analysis of non-normal data.
- These techniques are easy to understand and are often easier to apply. Thus unlike parametric procedures, the calculations with the complex distributions can be avoided.
- Although nonparametric procedures discard a portion of information about the data, but the loss in efficiency as compared to the competitors based on the normal parents is only nominal. However, such procedures tend to be more efficient when the underlying population deviates from normality.
- Nonparametric methods are robust in the sense that it tolerates departures from assumptions. That is, if the distributional assumption is perfect, one can adopt the usual parametric procedures. But, if the assumption of distribution is not perfect, nonparametric methods are the best alternative.

But nonparametric methods are not always the most desirable procedures as:

- Most of the nonparametric methods use only ranks or signs of the observations discarding the further features of the data. This makes such procedures less efficient.
- Nonparametric methods are usually not as efficient as their parametric counterparts when the assumptions are met.

Although nonparametric methods require fewer assumptions but they are used rarely in practice because of the use of limited information contained in the data. Consequently, nonparametric methods are primarily recommended in situations when assumptions are grossly violated (e.g. when severe skewness in the data is present).

Software based Learning

With the advent of high-speed computing facility and software like R, nonparametric techniques grow in a fast pace. We emphasize on using R for learning purpose. Actually R is open source statistical software and users can obtain the software free of charge through the Comprehensive R Archive Network (CRAN). Most importantly, if the required statistical methodology is not available within R, one can also write codes for his/her purpose. Consequently, for data based problems, we not only give the final result but also provide the R commands or codes together with the output.

An Exploratory Example of Real Field

Consider the following problem: Suppose the average weekly sales of a new cell phone are collected for 12 consecutive weeks from a particular shop are as below:

52 39 49 33 58 61 44 221 201 133 289 211

The average weekly sales were 125 units for the last year. Is the evidence sufficient to conclude that sales this year exceeds last year's sales?

The usual statistical procedure test in this context is a single sample t test. Specifically, if μ denotes the true weekly sales, then we are interested in testing H_0: $\mu = 125$ against H_1: $\mu > 125$. We run the test in R.

>data=c(52,39,49, 33, 58, 61, 44, 221, 133, 289, 211)

>t.test(data, mu=125, alternative="greater")

One Sample t-test

data: data

t = -0.6138, df = 10, p-value = 0.7235

alternative hypothesis: true mean is greater than 125

Thus we get the p value as.7235 and the value of the t statistic as -.6138 with 10 degrees of freedom. The tabulated value of a t statistic with 10 degrees of freedom is 1.81 at 5% level of significance. Thus we accept the null hypothesis at 5% significance level. Therefore, the evidence is not enough to reject the null hypothesis. Consequently, the evidence is not sufficient to conclude that there is an increase in the true mean weekly sales. But these results are not final because we have not yet checked the assumptions required to perform a t test. Specifically, results from t-tests are valid provided:

- Observations are drawn from a normal parent population.
- The sample size is sufficiently large (say, at least 30).

In our example, we have only 11 observations and hence t test is valid if we can show that the underlying distribution is normal. The simplest descriptive method to check the shape of the data is to plot the histogram. If the resulting histogram is symmetric and bell shaped, the underlying distribution is taken as normal. For the above data we plot the histogram below:

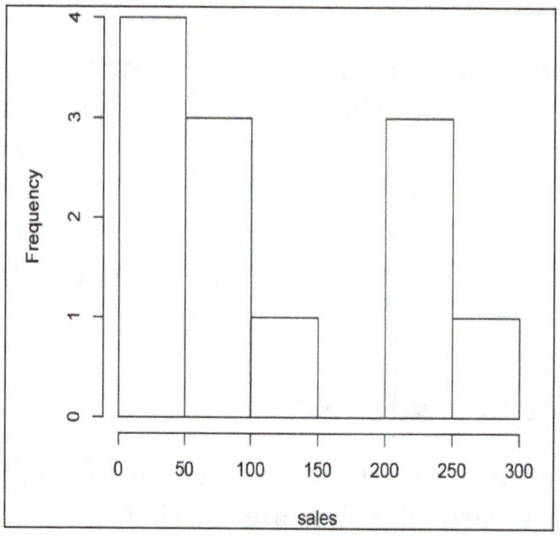

The histogram of data.

Clearly, the histogram of sales data is far from being symmetric. This indicates the non-normal nature of the underlying distribution of the sales data. However, for confirmation, we further use a normal Q-Q plot. Normal Q-Q plot is a simple and efficient exploratory technique, where the quantiles of the observed standardized data set are plotted against the corresponding quantiles of a standard normal distribution. If the underlying distribution is standard normal, the result should be a plot of points over a straight line. The normal Q–Q plot for the standardized version of the sales data is given below.

We find from the plot that the observed quintile's are too far from those of a standard normal distribution and hence the assumption of normality is not reasonable. Thus t test is not trustworthy for the sales data set.

This particular example exhibits the limitations of a parametric procedure in small samples, especially, when the underlying distributional assumption is not satisfied. The above data does not satisfy the assumptions required to perform a t test. As a result, here application of t test is forceful and hence the result from such a test is inconclusive even with a high p value.

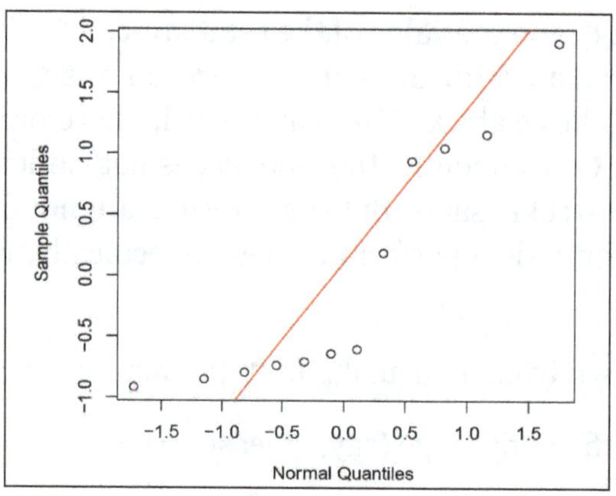

Normal Q–Q plot.

Thus there are situations in which parametric assumptions are violated and an alternative (i.e. nonparametric) method of analysis becomes the most appropriate and meaningful.

Point Estimation

In Statistics, point estimation is the process of finding an approximate value of some parameter—such as the mean (average)—of a population from random samples of the population. The accuracy of any particular approximation is not known precisely, though probabilistic statements concerning the accuracy of such numbers as found over many experiments can be constructed.

It is desirable for a point estimate to be: (1) Consistent. The larger the sample size, the more accurate the estimate. (2) Unbiased. The expectation of the observed values of many samples ("average observation value") equals the corresponding population parameter. For example, the sample mean is an unbiased estimator for the population mean. (3) Most efficient or best unbiased—of all consistent, unbiased estimates, the one possessing the smallest variance (a measure of the amount of dispersion away from the estimate). In other words, the estimator that varies least from sample to sample. This generally depends on the particular distribution of the population. For example, the mean is more efficient than the median (middle value) for the normal distribution but not for more "skewed" (asymmetrical) distributions.

Several methods are used to calculate the estimator. The most often used, the maximum likelihood method, uses differential calculus to determine the maximum of the probability function of a number of sample parameters. The moment's method equates values of sample moments (functions describing the parameter) to population moments. The solution of the equation gives the desired estimate. The Bayesian method, named for the 18th-century English theologian and mathematician Thomas Bayes, differs from the traditional methods by introducing a frequency function for the parameter being estimated. The drawback to the Bayesian method is that sufficient information on the distribution of the parameter is usually not available. One advantage is that the estimation can be easily adjusted as additional information becomes available.

Theory of Point Estimation

We consider a random experiment E. The outcome of E is represented by a Observable random vector $X = (X_1, X_2,..., X_n)$, $n \geq 1$. A particular value of X is denoted by $x = (x_1, x_2,..., x_n)$. The character X could be real or vector valued and the set of all values of X is called the sample space and it is denoted by $\mathcal{X} \subset \mathcal{R}^n$. The random vector X is generated by $F(x) = P(X \leq x)$, $x \in \mathcal{X}$, the distribution function of X.

In a parametric point estimation problem we assume that the functional form of F(x) is known except perhaps for a certain number of parameters. Let $\theta = (\theta_1, \theta_2,..., \theta_k)$ be the unknown parameters associated with F(x). The parameter θ may be real valued or vector valued and is usually called a labelling or indexing parameter. The labelling parameter θ varies over a set of values, called as parameter space and is denoted by $\Theta \subset \mathcal{R}^k$. So F(x) can be looked upon as a function of θ and henceforth we will write it as $F_\theta(x)$. If X is discrete or absolutely continuous then F(x) is generated by $f_\theta(x)$, the probability mass function (p.m.f.) or of probability density function (p.d.f.) X. We write

$\mathcal{F}_\theta = \{p(x, \theta):\theta \in \Theta\}$, as the class of all probability mass or density functions. The object of inference is the parameter θ or a function of the parameter θ say g(θ), which is of interest. Let us consider few examples. Example Suppose a coin is tossed 50 times.

Example: Suppose a coin is tossed 50 times. The outcome of *ith* toss can be described by a random variable X_i such that X_i = 1or 0 according as the *ith* toss results in a head or a tail.

$$\mathcal{X} = \{(x_1, x_2, \ldots x_{50}) : x_i = 0 \text{ or } 1 \text{ for all } i\}$$

If θ be the probability of getting a head in any toss then Θ = (0, 1) and the probability function of X is $p(x, \theta) = p(x,0) = \prod_{i=1}^{50} \theta^{x_i} (1-\theta)^{1-x_i}, x \in \mathcal{X}, \theta \in \Theta$. We may want to estimate θ or any function of θ.

Example: Suppose that 100 seeds of a certain flower were planted one in each pot and let X_i equal one or zero according as the seed in the *ith* pot germinates or not. The data consists of $(x_1, x_2, \ldots x_{100})$ a sequence of ones and zeroes and is regarded as a realization of $(X_1, X_2, \ldots X_{100})$ such that components are i.i.d. random variables with $P[X_1 = 1] = \theta$ and $P[X_1 = 0] = 1-\theta$, where θ represents the probability that a seed germinates. The object of estimation is θ itself or a function g(θ) that may be of interest. For example, consider $g(\theta) = \binom{10}{8}\theta^8 (1-\theta)^2$, which is the probability that in a batch of 10 seeds exactly 8 seeds will germinate.

Example: In a path-breaking experiment Rutherford Chadwick and Elis observed 2608 time intervals of 7.5 seconds each and counted the number of time intervals Nr in which exactly r number of α particles hit the counter. They obtained the following table:

r	0	1	2	3	4	5	6	7	8	9	≤10
N_r	57	203	383	525	532	408	273	139	45	27	18

It is quite well known that the Poisson distribution with p.m.f $f_\theta(x) = \dfrac{\exp(-\theta)\theta^x}{x!}, x = 1,2,\ldots \theta > 0$ serves as a good model for the number of times a given event E occurs in a unit time interval. If X_i denotes the number of α particles hitting the counter in the *i – th* time interval then $(X_1, X_2, \ldots X_n)$ where n = 2608 are i.i.d. Poisson random variables with parameter θ. We may want to estimate θ on the basis of the given data.

Example: Consider determination for an ideal physical constant such as gravity g. Usual way to estimate g is by the pendulum experiment and observe $X = \dfrac{4\pi^2 l}{T}$, , where l is the length of the pendulum and T the time required for a fixed number of oscillations. Due to variation which depends on several factors such as the skill of the experimenter and measurement errors, the *i – th* observation $X_i = g + e_i$ where e_i is the random error. Assuming the distribution of error is normal with zero mean and variance σ 2 we have $X_1, X_2, \ldots, X_n$ are i.i.d.N (g, σ²). Here the parameter θ is a two dimensional vector, θ = (g, σ²). Here we can view estimation of g. On the other hand one may be interested in estimating the error variance σ² through which we can estimate the ability of the experimenter.

Example: Suppose an experiment is conducted by measuring the length of lives in hours of n electric bulbs produced by a certain company. Let X_i be the length of live for the *ith* bulb.

$$\mathcal{X} = \{(x_1, x_2, \ldots x_n) : x_i \geq 0 \text{ for all } i\}$$

If we assume that the distribution of each X_i is exponential with mean θ then Θ = (0,∞) and the probability function of X is $p(x, \theta) = \prod_{i=1}^{n} \frac{1}{\theta} e^{-\frac{x_i}{\theta}}, x \in \mathcal{X}, \theta \in \Theta$. We may want to estimate the parameter θ or g(θ) = e – 60 θ, which represents the probability that the lifetime of a bulb will be at least 60 hours.

Objective

The distribution of X is characterized by the unknown parameter θ about which we only know that it belongs to the parameter space Θ. To discuss the problem of point estimation, for the sake of simplicity, we consider the case when the parameter of interest is a real valued function g = g(θ) of θ. In point estimation we try to approximate g(θ) on the basis of the observed value x of X. In other words we try to put forward a particular statistic or a function of X, say T = T(X), which would represent the unknown g(θ) very closely. Such a statistic T is called an estimator or a point estimator of g(θ). Mathematically, T is a measurable mapping from $\mathbb{X}$ to the space of g(θ) and it is called an admissible estimator. Any observed value of T is called an estimate of g(θ). In a nutshell, a point estimate of a parameter θ is a single number that can be regarded as a sensible value for θ. A point estimate is obtained by selecting a suitable statistic and computing its value from the given sample data. The selected statistic is called a point estimator of θ. It is to be noted that for a particular estimator T for a parameter θ the estimate of θ may vary from sample to sample.

Example: Suppose we want to estimate θ in Example of coin tossed 50 times. We may use the statistic $T = \frac{1}{n}\sum_{i=1}^{n} X_i$ as an estimator of θ. Here T is a mapping from $\mathbb{X}$ to (0, 1) and it is admissible.

Unbiased Estimation

An unbiased estimator is an accurate statistic that's used to approximate a population parameter. "Accurate" in this sense means that it's neither an overestimate nor an underestimate. If an over-estimate or underestimate does happen, the mean of the difference is called a "bias."

Estimable Function

A parametric function g(θ) is called estimable if there exists an at least one unbiased estimator of g(θ). Result: If X ~ Bin(n, θ) the estimable functions of θ are polynomial functions of θ of degree n or less. Proof: Let g(θ) be an estimable function of θ. Then there exists a statistic T(X) such that,

$$E_\theta\left(T(X)\right) = g(\theta) \text{ for all } \theta \in (0, 1)$$

$$\Rightarrow \sum_{i=0}^{n} T(x) \binom{n}{x} \theta^x (1-\theta)^{n-x} = g(\theta) \text{ for all } \theta \in (0, 1)$$

Since the L.H.S. of above equation is a polynomial in θ of degree n or less, to hold the identity the R.H.S. must be a polynomial in θ of degree n or less. Note: If X ~ Poisson(θ), the estimable functions of θ are convergent power series in θ.

Example: Consider a random variable X which takes only two values 1 or 2 with respective probabilities $\theta^2 + \theta^3$ and $1 - \theta^2 - \theta^3$. Here θ is an unknown parameter whose only possible values are 1/3 and 2/3. On the basis of a single observation X, we want to see whether the function $\frac{1}{\theta}$ is estimable or not. If possible suppose $\frac{1}{\theta}$ is estimable. Then there exists a statistic T(X) which is an unbiased estimator of $\frac{1}{\theta}$. Hence,

$$E_\theta(T)=\frac{1}{\theta},\ \text{for}\ \theta=1/3,2/3$$

$$\Rightarrow T(1)(\theta^2+\theta^3)+T(2)(1-\theta^2-\theta^3)=\frac{1}{\theta}\ \text{for}\ \theta=1/3,2/3$$

We get,

$$\frac{4}{27}T(1)+\frac{23}{27}T(2)=3$$

$$\frac{20}{27}T(1)+\frac{7}{27}T(2)=3/2$$

Two equations are satisfied when $T(1)=\frac{27}{32}$ and $T(2)=\frac{27}{8}$. Hence there exists an unbiased estimator of $\frac{1}{\theta}$ that is $\frac{1}{\theta}$ is estimable.

Example: Let $X_i \sim N(\mu, \sigma^2)$ i = 1, 2... n independently where μ and σ^2 are unknown.

Let $\theta = (\mu, \sigma^2)$ and $g(\theta) = e^{-\mu+\frac{1}{2}\sigma^2}$.

From the m.g.f. of the normal distribution $E_\theta(e^{tX}) = e^{t\mu+\frac{1}{2}t^2\sigma^2}$ for all θ and t.

Setting t = −1 we get $E_\theta(e^{-X}) = e^{-\mu+\frac{1}{2}\sigma^2}$ for all θ and hence,

$$E_\theta\left(\frac{1}{n}\sum_{i=1}^{n}e^{-X_i}\right)=e^{-\mu+\frac{1}{2}\sigma^2}\ \text{for all}\ \theta.$$

Thus there exists an unbiased estimator of $g(\theta)$ and hence $g(\theta)$ is estimable.

Best Linear Unbiased Estimator (BLUE)

The Best Linear Unbiased Estimate (BLUE) of a parameter θ based on data Y is:

- A linear function of Y. That is, the estimator can be written as b'Y.
- Unbiased (E [b'Y] = θ).
- Has the smallest variance among all unbiased linear estimators.

Linear combination c'θ, c'$\hat{Y}$ is the BLUE of c'θ, where $\hat{Y}$ is the least-squares orthogonal projection of Y onto $\mathcal{R}(X)$.

Corollary: If rank ($X_{n\times p}$) = p, then, for any a, a'$\hat{\beta}$ is the BLUE of a oβ. Note: The Gauss-Markov theorem generalizes this result to the less than full rank case, for certain linear combinations a'β (the estimable functions).

The Gauss-Markov theorem generalizes this result to the less than full rank case, for certain linear combinations a'β (the estimable functions).

Proof of Corollary:

$$\theta = X\beta$$
$$X'\theta = X'X\beta$$
$$(X'X)^{-1}X'\theta = (X'X)^{-1}X'X\beta = \beta$$
$$\Rightarrow a'\beta = \underbrace{a'(X'X)^{-1}X'}_{c'}\theta$$

So $a'\beta = c'\theta$ where $c' = a'(X'X)^{-1}X'$

Now $a'\beta = a'(X'X)^{-1}X'Y$ and

$$\begin{aligned} c'\hat{Y} &= a'(X'X)^{-1}X'\hat{Y} \\ &= a'(X'X)^{-1}X'X(X'X)^{-1}X'Y \\ &= a'(X'X)^{-1}X'Y \end{aligned}$$

Therefore, since $a'\hat{\beta} = c'\hat{Y}$, it is the BLUE of $a'\beta = c'\theta$.

4

Sampling and its Methods

The process of selecting a statistical sample which is a subset of individuals from a statistical population to determine the characteristics of the whole population is known as sampling. This chapter encompasses topics such as sampling, census, sampling frame, survey methodology and statistical unit. It closely determines the key concepts of sampling to provide an extensive understanding of the subject.

Sampling

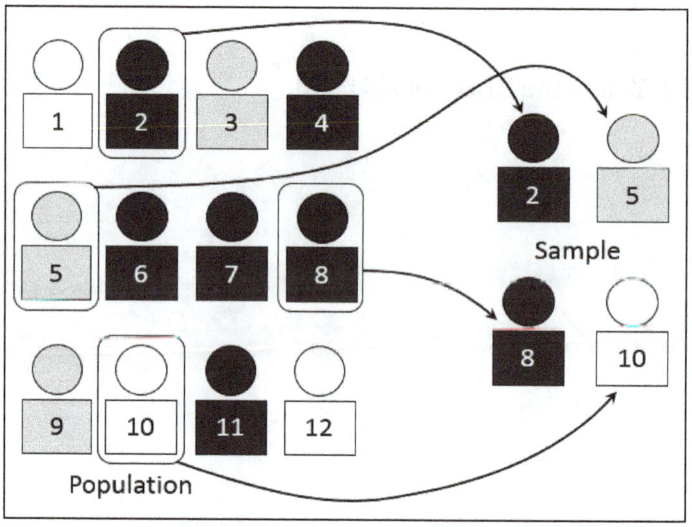

A visual representation of the sampling process.

In statistics, quality assurance, and survey methodology, sampling is concerned with the selection of a subset of individuals from within a statistical population to estimate characteristics of the whole population. Two advantages of sampling are that the cost is lower and data collection is faster than measuring the entire population.

Each observation measures one or more properties (such as weight, location, color) of observable bodies distinguished as independent objects or individuals. In survey sampling, weights can be applied to the data to adjust for the sample design, particularly stratified sampling. Results from probability theory and statistical theory are employed to guide the practice. In business and medical research, sampling is widely used for gathering information about a population. Acceptance sampling is used to determine if a production lot of material meets the governing specifications.

The sampling process comprises several stages:

- Defining the population of concern
- Specifying a sampling frame, a set of items or events possible to measure
- Specifying a sampling method for selecting items or events from the frame
- Determining the sample size
- Implementing the sampling plan
- Sampling and data collecting

Population Definition

Successful statistical practice is based on focused problem definition. In sampling, this includes defining the population from which our sample is drawn. A population can be defined as including all people or items with the characteristic one wishes to understand. Because there is very rarely enough time or money to gather information from everyone or everything in a population, the goal becomes finding a representative sample (or subset) of that population.

Sometimes what defines a population is obvious. For example, a manufacturer needs to decide whether a batch of material from production is of high enough quality to be released to the customer, or should be sentenced for scrap or rework due to poor quality. In this case, the batch is the population.

Although the population of interest often consists of physical objects, sometimes we need to sample over time, space, or some combination of these dimensions. For instance, an investigation of supermarket staffing could examine checkout line length at various times, or a study on endangered penguins might aim to understand their usage of various hunting grounds over time. For the time dimension, the focus may be on periods or discrete occasions.

In other cases, our 'population' may be even less tangible. For example, Joseph Jagger studied the behaviour of roulette wheels at a casino in Monte Carlo, and used this to identify a biased wheel. In this case, the 'population' Jagger wanted to investigate was the overall behaviour of the wheel (i.e. the probability distribution of its results over infinitely many trials), while his 'sample' was formed from observed results from that wheel. Similar considerations arise when taking repeated measurements of some physical characteristic such as the electrical conductivity of copper.

This situation often arises when we seek knowledge about the cause system of which the *observed* population is an outcome. In such cases, sampling theory may treat the observed population as a sample from a larger 'superpopulation'. For example, a researcher might study the success rate of a new 'quit smoking' program on a test group of 100 patients, in order to predict the effects of the program if it were made available nationwide. Here the superpopulation is "everybody in the country, given access to this treatment" – a group which does not yet exist, since the program isn't yet available to all.

Note also that the population from which the sample is drawn may not be the same as the population about which we actually want information. Often there is large but not complete overlap between these two groups due to frame issues etc. Sometimes they may be entirely separate – for instance, we might study rats in order to get a better understanding of human health, or we might study records from people born in 2008 in order to make predictions about people born in 2009.

Time spent in making the sampled population and population of concern precise is often well spent, because it raises many issues, ambiguities and questions that would otherwise have been overlooked at this stage.

Sampling Frame

In the most straightforward case, such as the sampling of a batch of material from production (acceptance sampling by lots), it would be most desirable to identify and measure every single item in the population and to include any one of them in our sample. However, in the more general case this is not usually possible or practical. There is no way to identify all rats in the set of all rats. Where voting is not compulsory, there is no way to identify which people will actually vote at a forthcoming election (in advance of the election). These imprecise populations are not amenable to sampling in any of the ways below and to which we could apply statistical theory.

As a remedy, we seek a sampling frame which has the property that we can identify every single element and include any in our sample. The most straightforward type of frame is a list of elements of the population (preferably the entire population) with appropriate contact information. For example, in an opinion poll, possible sampling frames include an electoral register and a telephone directory.

A probability sample is a sample in which every unit in the population has a chance (greater than zero) of being selected in the sample, and this probability can be accurately determined. The combination of these traits makes it possible to produce unbiased estimates of population totals, by weighting sampled units according to their probability of selection.

Example: We want to estimate the total income of adults living in a given street. We visit each household in that street, identify all adults living there, and randomly select one adult from each household. (For example, we can allocate each person a random number, generated from a uniform distribution between 0 and 1, and select the person with the highest number in each household). We then interview the selected person and find their income.

People living on their own are certain to be selected, so we simply add their income to our estimate of the total. But a person living in a household of two adults has only a one-in-two chance of selection. To reflect this, when we come to such a household, we would count the selected person's income twice towards the total. (The person who is *selected from that household can be loosely viewed as also representing the person who* isn't *selected).*

In the above example, not everybody has the same probability of selection; what makes it a probability sample is the fact that each person's probability is known. When every element in the population *does* have the same probability of selection, this is known as an 'equal probability of selection' (EPS) design. Such designs are also referred to as 'self-weighting' because all sampled units are given the same weight.

Probability sampling includes: Simple Random Sampling, Systematic Sampling, Stratified Sampling, Probability Proportional to Size Sampling, and Cluster or Multistage Sampling. These various ways of probability sampling have two things in common:

1. Every element has a known nonzero probability of being sampled and
2. Involves random selection at some point.

Nonprobability Sampling

Nonprobability sampling is any sampling method where some elements of the population have *no* chance of selection (these are sometimes referred to as 'out of coverage'/'undercovered'), or where the probability of selection can't be accurately determined. It involves the selection of elements based on assumptions regarding the population of interest, which forms the criteria for selection. Hence, because the selection of elements is nonrandom, nonprobability sampling does not allow the estimation of sampling errors. These conditions give rise to exclusion bias, placing limits on how much information a sample can provide about the population. Information about the relationship between sample and population is limited, making it difficult to extrapolate from the sample to the population.

Example: We visit every household in a given street, and interview the first person to answer the door. In any household with more than one occupant, this is a nonprobability sample, because some people are more likely to answer the door (e.g. an unemployed person who spends most of their time at home is more likely to answer than an employed housemate who might be at work when the interviewer calls) and it's not practical to calculate these probabilities.

Nonprobability sampling methods include convenience sampling, quota sampling and purposive sampling. In addition, nonresponse effects may turn *any* probability design into a nonprobability design if the characteristics of nonresponse are not well understood, since nonresponse effectively modifies each element's probability of being sampled.

Sampling Methods

Within any of the types of frame identified above, a variety of sampling methods can be employed, individually or in combination. Factors commonly influencing the choice between these designs include:

- Nature and quality of the frame
- Availability of auxiliary information about units on the frame
- Accuracy requirements, and the need to measure accuracy
- Whether detailed analysis of the sample is expected
- Cost/operational concerns

Simple Random Sampling

In a simple random sample (SRS) of a given size, all such subsets of the frame are given an equal probability. Furthermore, any given *pair* of elements has the same chance of selection as any other such pair (and similarly for triples, and so on). This minimizes bias and simplifies analysis of results. In particular, the variance between individual results within the sample is a good indicator of variance in the overall population, which makes it relatively easy to estimate the accuracy of results.

SRS can be vulnerable to sampling error because the randomness of the selection may result in a sample that doesn't reflect the makeup of the population. For instance, a simple random sample of ten people from a given country will *on average* produce five men and five women, but any given trial is likely to overrepresent one sex and underrepresent the other. Systematic and stratified techniques attempt to overcome this problem by "using information about the population" to choose a more "representative" sample.

SRS may also be cumbersome and tedious when sampling from an unusually large target population. In some cases, investigators are interested in "research questions specific" to subgroups of the population. For example, researchers might be interested in examining whether cognitive ability as a predictor of job performance is equally applicable across racial groups. SRS cannot accommodate the needs of researchers in this situation because it does not provide subsamples of the population. "Stratified sampling" addresses this weakness of SRS.

Systematic Sampling

Systematic sampling (also known as interval sampling) relies on arranging the study population according to some ordering scheme and then selecting elements at regular intervals through that ordered list. Systematic sampling involves a random start and then proceeds with the selection of every *k*th element from then onwards. In this case, *k*=(population size/sample size). It is important that the starting point is not automatically the first in the list, but is instead randomly chosen from within the first to the *k*th element in the list. A simple example would be to select every 10th name from the telephone directory (an 'every 10th' sample, also referred to as 'sampling with a skip of 10').

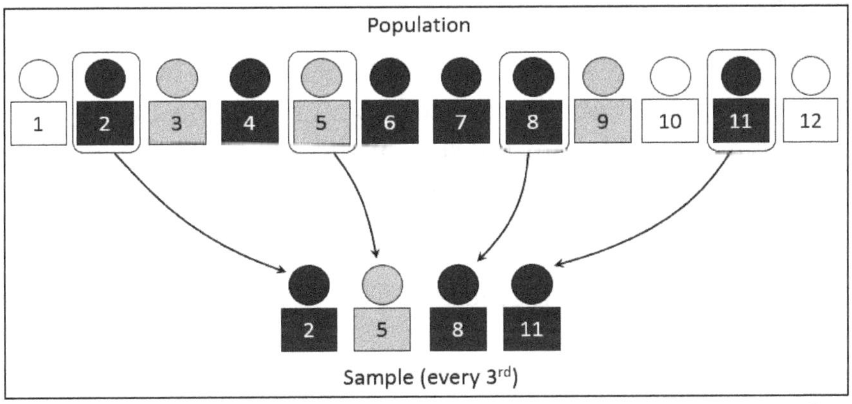

A visual representation of selecting a random sample using the systematic sampling technique.

As long as the starting point is randomized, systematic sampling is a type of probability sampling. It is easy to implement and the stratification induced can make it efficient, *if* the variable by which the list is ordered is correlated with the variable of interest. 'Every 10th' sampling is especially useful for efficient sampling from databases.

For example, suppose we wish to sample people from a long street that starts in a poor area (house No. 1) and ends in an expensive district (house No. 1000). A simple random selection of addresses from this street could easily end up with too many from the high end and too few from the low end (or vice versa), leading to an unrepresentative sample. Selecting (e.g.) every 10th street number along the street ensures that the sample is spread evenly along the length of the street, representing all of these districts. (Note that if we always start at house #1 and end at #991, the sample is slightly biased towards the low end; by randomly selecting the start between #1 and #10, this bias is eliminated.

However, systematic sampling is especially vulnerable to periodicities in the list. If periodicity is present and the period is a multiple or factor of the interval used, the sample is especially likely to be *un*representative of the overall population, making the scheme less accurate than simple random sampling.

For example, consider a street where the odd-numbered houses are all on the north (expensive) side of the road, and the even-numbered houses are all on the south (cheap) side. Under the sampling scheme given above, it is impossible to get a representative sample; either the houses sampled will *all* be from the odd-numbered, expensive side, or they will *all* be from the even-numbered, cheap side, unless the researcher has previous knowledge of this bias and avoids it by a using a skip which ensures jumping between the two sides (any odd-numbered skip).

Another drawback of systematic sampling is that even in scenarios where it is more accurate than SRS, its theoretical properties make it difficult to *quantify* that accuracy. (In the two examples of systematic sampling that are given above, much of the potential sampling error is due to variation between neighbouring houses – but because this method never selects two neighbouring houses, the sample will not give us any information on that variation).

As described above, systematic sampling is an EPS method, because all elements have the same probability of selection (in the example given, one in ten). It is *not* 'simple random sampling' because different subsets of the same size have different selection probabilities – e.g. the set {4,14,24,...,994} has a one-in-ten probability of selection, but the set {4,13,24,34,...} has zero probability of selection.

Systematic sampling can also be adapted to a non-EPS approach; for an example.

Stratified Sampling

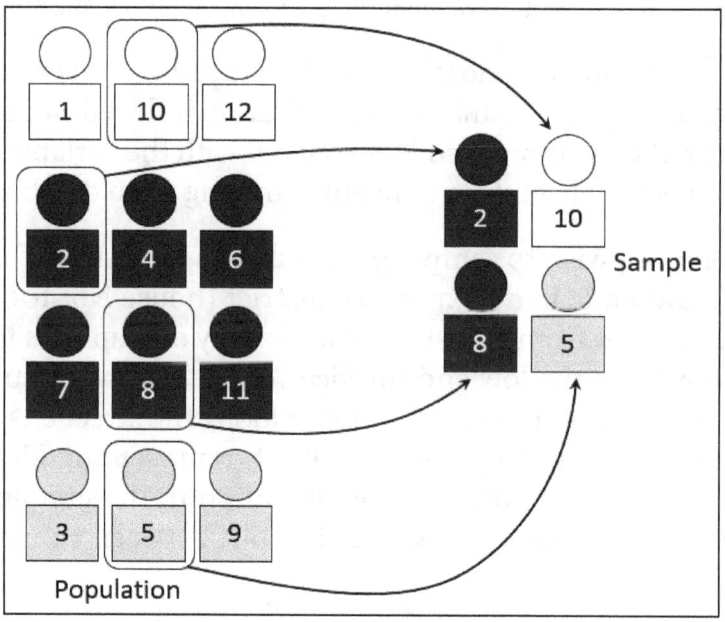

A visual representation of selecting a random sample using the stratified sampling technique.

When the population embraces a number of distinct categories, the frame can be organized by these categories into separate "strata." Each stratum is then sampled as an independent sub-population, out of which individual elements can be randomly selected. There are several potential benefits to stratified sampling.

First, dividing the population into distinct, independent strata can enable researchers to draw inferences about specific subgroups that may be lost in a more generalized random sample.

Second, utilizing a stratified sampling method can lead to more efficient statistical estimates (provided that strata are selected based upon relevance to the criterion in question, instead of availability of the samples). Even if a stratified sampling approach does not lead to increased statistical efficiency, such a tactic will not result in less efficiency than would simple random sampling, provided that each stratum is proportional to the group's size in the population.

Third, it is sometimes the case that data are more readily available for individual, pre-existing strata within a population than for the overall population; in such cases, using a stratified sampling approach may be more convenient than aggregating data across groups (though this may potentially be at odds with the previously noted importance of utilizing criterion-relevant strata).

Finally, since each stratum is treated as an independent population, different sampling approaches can be applied to different strata, potentially enabling researchers to use the approach best suited (or most cost-effective) for each identified subgroup within the population.

There are, however, some potential drawbacks to using stratified sampling. First, identifying strata and implementing such an approach can increase the cost and complexity of sample selection, as well as leading to increased complexity of population estimates. Second, when examining multiple criteria, stratifying variables may be related to some, but not to others, further complicating the design, and potentially reducing the utility of the strata. Finally, in some cases (such as designs with a large number of strata, or those with a specified minimum sample size per group), stratified sampling can potentially require a larger sample than would other methods (although in most cases, the required sample size would be no larger than would be required for simple random sampling.

A stratified sampling approach is most effective when three conditions are met:

1. Variability within strata are minimized.
2. Variability between strata are maximized.
3. The variables upon which the population is stratified are strongly correlated with the desired dependent variable.

Advantages over other sampling methods:

1. Focuses on important subpopulations and ignores irrelevant ones.
2. Allows use of different sampling techniques for different subpopulations.
3. Improves the accuracy/efficiency of estimation.
4. Permits greater balancing of statistical power of tests of differences between strata by sampling equal numbers from strata varying widely in size.

Disadvantages:

1. Requires selection of relevant stratification variables which can be difficult.
2. Is not useful when there are no homogeneous subgroups.
3. Can be expensive to implement.

Poststratification

Stratification is sometimes introduced after the sampling phase in a process called "post-stratification". This approach is typically implemented due to a lack of prior knowledge of an appropriate stratifying variable or when the experimenter lacks the necessary information to create a stratifying variable during the sampling phase. Although the method is susceptible to the pitfalls of post hoc approaches, it can provide several benefits in the right situation. Implementation usually follows a simple random sample. In addition to allowing for stratification on an ancillary variable, poststratification can be used to implement weighting, which can improve the precision of a sample's estimates.

Oversampling

Choice-based sampling is one of the stratified sampling strategies. In choice-based sampling, the data are stratified on the target and a sample is taken from each stratum so that the rare target class will be more represented in the sample. The model is then built on this biased sample. The effects of the input variables on the target are often estimated with more precision with the choice-based sample even when a smaller overall sample size is taken, compared to a random sample. The results usually must be adjusted to correct for the oversampling.

Probability-proportional-to-size Sampling

In some cases the sample designer has access to an "auxiliary variable" or "size measure", believed to be correlated to the variable of interest, for each element in the population. These data can be used to improve accuracy in sample design. One option is to use the auxiliary variable as a basis for stratification.

Another option is probability proportional to size ('PPS') sampling, in which the selection probability for each element is set to be proportional to its size measure, up to a maximum of 1. In a simple PPS design, these selection probabilities can then be used as the basis for Poisson sampling. However, this has the drawback of variable sample size, and different portions of the population may still be over- or under-represented due to chance variation in selections.

Systematic sampling theory can be used to create a probability proportionate to size sample. This is done by treating each count within the size variable as a single sampling unit. Samples are then identified by selecting at even intervals among these counts within the size variable. This method is sometimes called PPS-sequential or monetary unit sampling in the case of audits or forensic sampling.

Example: Suppose we have six schools with populations of 150, 180, 200, 220, 260, and 490 students respectively (total 1500 students), and we want to use student population as the basis for a PPS sample of size three. To do this, we could allocate the first school numbers 1 to 150, the second school 151 to 330 (= 150 + 180), the third school 331 to 530, and so on to the last school (1011 to 1500). We then generate a random start between 1 and 500 (equal to 1500/3) and count through the school populations by multiples of 500. If our random start was 137, we would select the schools which have been allocated numbers 137, 637, and 1137, i.e. the first, fourth, and sixth schools.

The PPS approach can improve accuracy for a given sample size by concentrating sample on large elements that have the greatest impact on population estimates. PPS sampling is commonly used for surveys of businesses, where element size varies greatly and auxiliary information is often available—for instance, a survey attempting to measure the number of guest-nights spent in hotels might use each hotel's number of rooms as an auxiliary variable. In some cases, an older measurement of the variable of interest can be used as an auxiliary variable when attempting to produce more current estimates.

Cluster Sampling

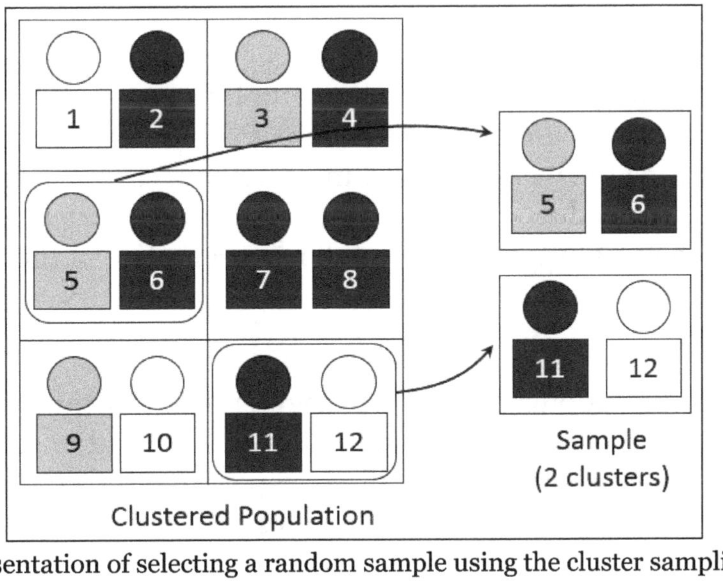

A visual representation of selecting a random sample using the cluster sampling technique.

Sometimes it is more cost-effective to select respondents in groups ('clusters'). Sampling is often clustered by geography, or by time periods. (Nearly all samples are in some sense 'clustered' in time – although this is rarely taken into account in the analysis.) For instance, if surveying households within a city, we might choose to select 100 city blocks and then interview every household within the selected blocks.

Clustering can reduce travel and administrative costs. An interviewer can make a single trip to visit several households in one block, rather than having to drive to a different block for each household.

It also means that one does not need a sampling frame listing all elements in the target population. Instead, clusters can be chosen from a cluster-level frame, with an element-level frame created only for the selected clusters. The sample only requires a block-level city map for initial selections, and then a household-level map of the 100 selected blocks, rather than a household-level map of the whole city.

Cluster sampling (also known as clustered sampling) generally increases the variability of sample estimates above that of simple random sampling, depending on how the clusters differ between one another as compared to the within-cluster variation. For this reason, cluster sampling requires a larger sample than SRS to achieve the same level of accuracy – but cost savings from clustering might still make this a cheaper option.

Cluster sampling is commonly implemented as multistage sampling. This is a complex form of cluster sampling in which two or more levels of units are embedded one in the other. The first stage consists of constructing the clusters that will be used to sample from. In the second stage, a sample of primary units is randomly selected from each cluster (rather than using all units contained in all selected clusters). In following stages, in each of those selected clusters, additional samples of units are selected, and so on. All ultimate units (individuals, for instance) selected at the last step of this procedure are then surveyed. This technique, thus, is essentially the process of taking random subsamples of preceding random samples.

Multistage sampling can substantially reduce sampling costs, where the complete population list would need to be constructed (before other sampling methods could be applied). By eliminating the work involved in describing clusters that are not selected, multistage sampling can reduce the large costs associated with traditional cluster sampling. However, each sample may not be a full representative of the whole population.

Quota Sampling

In quota sampling, the population is first segmented into mutually exclusive subgroups, just as in stratified sampling. Then judgement is used to select the subjects or units from each segment based on a specified proportion. For example, an interviewer may be told to sample 200 females and 300 males between the age of 45 and 60.

It is this second step which makes the technique one of non-probability sampling. In quota sampling the selection of the sample is non-random. For example, interviewers might be tempted to interview those who look most helpful. The problem is that these samples may be biased because not everyone gets a chance of selection. This random element is its greatest weakness and quota versus probability has been a matter of controversy for several years.

Minimax Sampling

In imbalanced datasets, where the sampling ratio does not follow the population statistics, one can resample the dataset in a conservative manner called minimax sampling. The minimax sampling has its origin in Anderson minimax ratio whose value is proved to be 0.5: in a binary classification, the class-sample sizes should be chosen equally. This ratio can be proved to be minimax ratio only under the assumption of LDA classifier with Gaussian distributions. The notion of minimax sampling is recently developed for a general class of classification rules, called class-wise smart classifiers.

Accidental Sampling

Accidental sampling (sometimes known as grab, convenience or opportunity sampling) is a type of nonprobability sampling which involves the sample being drawn from that part of the population which is close to hand. That is, a population is selected because it is readily available and convenient. It may be through meeting the person or including a person in the sample when one meets them or chosen by finding them through technological means such as the internet or through phone. The researcher using such a sample cannot scientifically make generalizations about the total population from this sample because it would not be representative enough. For example, if the interviewer were to conduct such a survey at a shopping center early in the morning on a given day, the people that he/she could interview would be limited to those given there at that given time, which would not represent the views of other members of society in such an area, if the survey were to be conducted at different times of day and several times per week. This type of sampling is most useful for pilot testing. Several important considerations for researchers using convenience samples include:

1. Are there controls within the research design or experiment which can serve to lessen the impact of a non-random convenience sample, thereby ensuring the results will be more representative of the population?

2. Is there good reason to believe that a particular convenience sample would or should respond or behave differently than a random sample from the same population?

3. Is the question being asked by the research one that can adequately be answered using a convenience sample?

In social science research, snowball sampling is a similar technique, where existing study subjects are used to recruit more subjects into the sample. Some variants of snowball sampling, such as respondent driven sampling, allow calculation of selection probabilities and are probability sampling methods under certain conditions.

Line-intercept Sampling

Line-intercept sampling is a method of sampling elements in a region whereby an element is sampled if a chosen line segment, called a "transect", intersects the element.

Panel Sampling

Panel sampling is the method of first selecting a group of participants through a random sampling method and then asking that group for (potentially the same) information several times over a period of time. Therefore, each participant is interviewed at two or more time points; each period of data collection is called a "wave". The method was developed by sociologist Paul Lazarsfeld in 1938 as a means of studying political campaigns. This longitudinal sampling-method allows estimates of changes in the population, for example with regard to chronic illness to job stress to weekly food expenditures. Panel sampling can also be used to inform researchers about within-person health changes due to age or to help explain changes in continuous dependent variables such as spousal interaction. There have been several proposed methods of analyzing panel data, including MANOVA, growth curves, and structural equation modeling with lagged effects.

Snowball Sampling

Snowball sampling involves finding a small group of initial respondents and using them to recruit more respondents. It is particularly useful in cases where the population is hidden or difficult to enumerate.

Theoretical Sampling

Theoretical sampling occurs when samples are selected on the basis of the results of the data collected so far with a goal of developing a deeper understanding of the area or develop theory.

Replacement of Selected Units

Sampling schemes may be *without replacement* ('WOR'—no element can be selected more than once in the same sample) or *with replacement* ('WR'—an element may appear multiple times in the one sample). For example, if we catch fish, measure them, and immediately return them to the water before continuing with the sample, this is a WR design, because we might end up catching and measuring the same fish more than once. However, if we do not return the fish to the water (e.g., if we eat the fish), this becomes a WOR design.

Sample Sizc

Formulas, tables, and power function charts are well known approaches to determine sample size.

Steps for Using Sample Size Tables

1. Postulate the effect size of interest, α, and β.
2. Check sample size table:
   - Select the table corresponding to the selected α.
   - Locate the row corresponding to the desired power.
   - Locate the column corresponding to the estimated effect size.
   - The intersection of the column and row is the minimum sample size required.

Sampling and Data Collection

Good data collection involves:

- Following the defined sampling process
- Keeping the data in time order
- Noting comments and other contextual events
- Recording non-responses

Applications of Sampling

Sampling enables the selection of right data points from within the larger data set to estimate the characteristics of the whole population. For example, there are about 600 million tweets produced every day. It is not necessary to look at all of them to determine the topics that are discussed during the day, nor is it necessary to look at all the tweets to determine the sentiment on each of the topics. A theoretical formulation for sampling Twitter data has been developed.

In manufacturing different types of sensory data such as acoustics, vibration, pressure, current, voltage and controller data are available at short time intervals. To predict down-time it may not be necessary to look at all the data but a sample may be sufficient.

Errors in Sample Surveys

Survey results are typically subject to some error. Total errors can be classified into sampling errors and non-sampling errors. The term "error" here includes systematic biases as well as random errors.

Sampling Errors and Biases

Sampling errors and biases are induced by the sample design. They include:

1. Selection bias: When the true selection probabilities differ from those assumed in calculating the results.
2. Random sampling error: Random variation in the results due to the elements in the sample being selected at random.

Non-sampling Error

Non-sampling errors are other errors which can impact the final survey estimates, caused by problems in data collection, processing, or sample design. They include:

1. Over-coverage: Inclusion of data from outside of the population.
2. Under-coverage: Sampling frame does not include elements in the population.
3. Measurement error: e.g. when respondents misunderstand a question, or find it difficult to answer.
4. Processing error: Mistakes in data coding.
5. Non-response: Failure to obtain complete data from all selected individuals.

After sampling, a review should be held of the exact process followed in sampling, rather than that intended, in order to study any effects that any divergences might have on subsequent analysis. A particular problem is that of *non-response*. Two major types of non-response exist: unit nonresponse (referring to lack of completion of any part of the survey) and item non-response (submission or participation in survey but failing to complete one or more components/questions of the survey). In survey sampling, many of the individuals identified as part of the sample may be unwilling to participate, not have the time to participate (opportunity cost), or survey administrators may not have been able to contact them. In this case, there is a risk of differences, between respondents and nonrespondents, leading to biased estimates of population parameters. This is often addressed by improving survey design, offering incentives, and conducting follow-up studies which make a repeated attempt to contact the unresponsive and to characterize their similarities and differences with the rest of the frame. The effects can also be mitigated by weighting the data when population benchmarks are available or by imputing data based on answers to other questions.

Nonresponse is particularly a problem in internet sampling. Reasons for this problem include improperly designed surveys, over-surveying (or survey fatigue), and the fact that potential participants hold multiple e-mail addresses, which they don't use anymore or don't check regularly.

Survey Weights

In many situations the sample fraction may be varied by stratum and data will have to be weighted to correctly represent the population. Thus for example, a simple random sample of individuals in the United Kingdom might include some in remote Scottish islands who would be inordinately expensive to sample. A cheaper method would be to use a stratified sample with urban and rural strata. The rural sample could be under-represented in the sample, but weighted up appropriately in the analysis to compensate.

More generally, data should usually be weighted if the sample design does not give each individual an equal chance of being selected. For instance, when households have equal selection probabilities but one person is interviewed from within each household, this gives people from large households a smaller chance of being interviewed. This can be accounted for using survey weights. Similarly, households with more than one telephone line have a greater chance of being selected in a random digit dialing sample, and weights can adjust for this.

Weights can also serve other purposes, such as helping to correct for non-response.

Methods of Producing Random Samples

- Random number table
- Mathematical algorithms for pseudo-random number generators
- Physical randomization devices such as coins, playing cards or sophisticated devices such as ERNIE

History

Random sampling by using lots is an old idea, mentioned several times in the Bible. In 1786 Pierre Simon Laplace estimated the population of France by using a sample, along with ratio estimator. He also computed probabilistic estimates of the error. These were not expressed as modern confidence intervals but as the sample size that would be needed to achieve a particular upper bound on the sampling error with probability 1000/1001. His estimates used Bayes' theorem with a uniform prior probability and assumed that his sample was random. Alexander Ivanovich Chuprov introduced sample surveys to Imperial Russia in the 1870s.

In the USA the 1936 *Literary Digest* prediction of a Republican win in the presidential election went badly awry, due to severe bias. More than two million people responded to the study with their names obtained through magazine subscription lists and telephone directories. It was not appreciated that these lists were heavily biased towards Republicans and the resulting sample, though very large, was deeply flawed.

Statistics is the Science of Data

Data are numerical values containing some information.

Statistical tools can be used on a data set to draw statistical inferences. These statistical inferences are in turn used for various purposes. For example, government uses such data for policy formulation for the welfare of the people, marketing companies use the data from consumer surveys to improve the company and to provide better services to the customer, etc. Such data is obtained through sample surveys. Sample surveys are conducted throughout the world by governmental as well as non-governmental agencies. For example, "National Sample Survey Organization (NSSO)" conducts surveys in India, "Statistics Canada" conducts surveys in Canada, agencies of United Nations like "World Health Organization (WHO), "Food and Agricultural Organization (FAO)" etc. conduct surveys in different countries.

Sampling theory provides the tools and techniques for data collection keeping in mind the objectives to be fulfilled and nature of population.

There are two ways of obtaining the information

1. Sample surveys
2. Complete enumeration or census

Sample surveys collect information on a fraction of total population whereas in census, the information is collected on the whole population. Some surveys, e.g. economic surveys, agricultural surveys etc. are conducted regularly. Some surveys are need based and are conducted when some need arises, e.g., consumer satisfaction surveys at a newly opened shopping mall to see the satisfaction level with the amenities provided in the mall.

Sampling Unit

An element or a group of elements on which observations can be taken is called a sampling unit. The objective of the survey helps in determining the definition of sampling unit.

For example, if the objective is to determine the total income of all the persons in the household, then the sampling unit is household. If the objective is to determine the income of any particular person in the household, then the sampling unit is the income of the particular person in the household. So the definition of sampling unit depends and varies as per the objective of the survey. Similarly, in another example, if the objective is to study the blood sugar level, then the sampling unit is the value of blood sugar level of a person. On the other hand, if the objective is to study the health conditions, then the sampling unit is the person on whom the readings on the blood sugar level, blood pressure and other factors will be obtained. These values will together classify the person as healthy or unhealthy.

Population

Collection of all the sampling units in a given region at a particular point of time or a particular period is called population. For example, if the medical facilities in a hospital are to be surveyed through the patients, then the total number of patients registered in the hospital during the time period of survey will be the population. Similarly, if the production of wheat in a district is to be studied, then all the fields cultivating wheat in that district will constitute the population. The total number of sampling units in the population is the population size, denoted generally by N. The population size can be finite or infinite (N is large).

Census

Census taker visits a Romani family living in a caravan, Netherlands 1925.

A census is the procedure of systematically acquiring and recording information about the members of a given population. It is a regularly occurring and official count of a particular population. The term is used mostly in connection with national population and housing censuses; other common censuses include agriculture, business, and traffic censuses. The United Nations defines the essential features of population and housing censuses as "individual enumeration, universality within a defined territory, simultaneity and defined periodicity", and recommends that population censuses be taken at least every 10 years. United Nations recommendations also cover census topics to be collected, official definitions, classifications and other useful information to co-ordinate international practice.

The word is of Latin origin: during the Roman Republic, the census was a list that kept track of all adult males fit for military service. The modern census is essential to international comparisons of any kind of statistics, and censuses collect data on many attributes of a population, not just how many people there are but now census takes its place within a system of surveys where it typically began as the only national demographic data collection. Although population estimates remain an important function of a census, including exactly the geographic distribution of the population, statistics can be produced about combinations of attributes e.g. education by age and sex in different regions. Current administrative data systems allow for other approaches to enumeration with the same level of detail but raise concerns about privacy and the possibility of biasing estimates.

A census can be contrasted with sampling in which information is obtained only from a subset of a population, typically main population estimates are updated by such intercensal estimates. Modern census data are commonly used for research, business marketing, and planning, and as a baseline for designing sample surveys by providing a sampling frame such as an address register. Census counts are necessary to adjust samples to be representative of a population by weighting them as is common in opinion polling. Similarly, stratification requires knowledge of the relative sizes of different population strata which can be derived from census enumerations. In some countries, the census provides the official counts used to apportion the number of elected representatives to regions (sometimes controversially – e.g., *Utah v. Evans*). In many cases, a carefully chosen random sample can provide more accurate information than attempts to get a population census.

Sampling

A census is often construed as the opposite of a sample as its intent is to count everyone in a population rather than a fraction. However, population censuses rely on a sampling frame to count the population. This is the only way to be sure that everyone has been included as otherwise those not responding would not be followed up on and individuals could be missed. The fundamental premise of a census is that the population is not known and a new estimate is to be made by the analysis of primary data. The use of a sampling frame is counterintuitive as it suggests that the population size is already known. However, a census is also used to collect attribute data on the individuals in the nation. This process of sampling marks the difference between historical census, which was a house to house process or the product of an imperial decree, and the modern statistical project. The sampling frame used by census is almost always an address register. Thus it is not known if there is anyone resident or how many people there are in each household. Depending on the mode of enumeration, a form is sent to the householder, an enumerator calls, or administrative records for the dwelling are accessed. As a preliminary to the dispatch of forms, census workers will check any address problems on the ground. While it may seem straightforward to use the postal service file for this purpose, this can be out of date and some dwellings may contain a number of independent households. A particular problem is what are termed 'communal establishments' which category includes student residences, religious orders, homes for the elderly, people in prisons etc. As these are not easily enumerated by a single householder, they are often treated differently and visited by special teams of census workers to ensure they are classified appropriately.

Residence Definitions

Individuals are normally counted within households and information is typically collected about the household structure and the housing. For this reason international documents refer to censuses of population and housing. Normally the census response is made by a household, indicating details of individuals resident there. An important aspect of census enumerations is determining which individuals can be counted from which cannot be counted. Broadly, three definitions can be used: *de facto* residence; *de jure* residence; and, permanent residence. This is important to consider individuals who have multiple or temporary addresses. Every person should be identified uniquely as resident in one place but where they happen to be on Census Day, their *de facto* residence, may not be the best place to count them. Where an individual uses services may be more useful and this is at their usual, or *de jure*, residence. An individual may be represented at a permanent address, perhaps a family home for students or long term migrants. It is necessary to have a precise definition of residence to decide whether visitors to a country should be included in the population count. This is becoming more important as students travel abroad for education for a period of several years. Other groups causing problems of enumeration are new born babies, refugees, people away on holiday, people moving home around census day, and people without a fixed address. People having second homes because of working in another part of the country or retaining a holiday cottage are difficult to fix at a particular address sometimes causing double counting or houses being mistakenly identified as vacant. Another problem is where people use a different address at different times e.g. students living at their place of education in term time but returning to a family home during vacations or children whose parents have separated who effectively have two family homes. Census enumeration has always been based on finding people where they live as there is no systematic alternative - any list you could use to find people is derived from census activities in the first place. Recent UN guidelines provide recommendation on enumerating such complex households.

Enumeration Strategies

Historical censuses used crude enumeration assuming absolute accuracy. Modern approaches take into account the problems of overcount and undercount, and the coherence of census enumerations with other official sources of data. This reflects a realist approach to measurement, acknowledging that under any definition of residence there is a true value of the population but this can never be measured with complete accuracy. An important aspect of the census process is to evaluate the quality of the data.

Many countries use a post-enumeration survey to adjust the raw census counts. This works in a similar manner to capture-recapture estimation for animal populations. In census circles this method is called dual system enumeration (DSE). A sample of households are visited by interviewers who record the details of the household as at census day. These data are then matched to census records and the number of people missed can be estimated by considering the number missed in the census or survey but counted in the other. This way counts can be adjusted for non-response varying between different demographic groups. An explanation using a fishing analogy can be found in "Trout, Catfish and Roach..." which won an award from the Royal Statistical Society for excellence in official statistics in 2011.

Triple system enumeration has been proposed as an improvement as it would allow evaluation of the statistical dependence of pairs of sources. However, as the matching process is the most difficult aspect of census estimation this has never been implemented for a national enumeration. It would also be difficult to identify three different sources that were sufficiently different to make the triple system effort worthwhile. The DSE approach has another weakness in that it assumes there is no person counted twice (over count). In *de facto* residence definitions this would not be a problem but in *de jure* definitions individuals risk being recorded on more than one form leading to double counting. A particular problem here are students who often have a term time and family address.

Several countries have used a system which is known as short form/long form. This is a sampling strategy which randomly chooses a proportion of people to send a more detailed questionnaire to (the long form). Everyone receives the short form questions. Thereby more data are collected but not imposing a burden on the whole population. This also reduces the burden on the statistical office. Indeed, in the UK all residents were required to fill in the whole form but only a 10% sample were coded and analysed in detail, until 2001. New technology means that all data are now scanned and processed. Recently there has been controversy in Canada about the cessation of the long form with the head, Munir Sheikh resigning.

The use of alternative enumeration strategies is increasing but these are not so simple as many people assume and only occur in developed countries. The Netherlands has been most advanced in adopting a census using administrative data. This allows a simulated census to be conducted by linking several different administrative databases at an agreed time. Data can be matched and an overall enumeration established accounting for where the different sources are discrepant. A validation survey is still conducted in a similar way to the post enumeration survey employed in a traditional census. Other countries which have a population register use this as a basis for all the census statistics needed by users. This is most common amongst Nordic countries but requires a large number of different registers to be combined including population, housing, employment and education. These registers are then combined and brought up to the standard of a statistical register by comparing the data in different sources and ensuring the quality is sufficient for official statistics to be produced. A recent innovation is the French instigation of a rolling census programme with different regions enumerated each year such that the whole country is completely enumerated every 5 to 10 years. In Europe, in connection with the 2010 census round, a large number of countries adopted alternative census methodologies, often based on the combination of data from registers, surveys and other sources.

Technology

Censuses have evolved in their use of technology with the latest censuses, the 2010 round, using many new types of computing. In Brazil, handheld devices were used by enumerators to locate residences on the ground. In many countries, census returns could be made via the Internet as well as in paper form. DSE is facilitated by computer matching techniques which can be automated, such as propensity score matching. In the UK, all census formats are scanned and stored electronically before being destroyed, replacing the need for physical archives. The record linking to perform an administrative census would not be possible without large databases being stored on computer systems.

New technology is not without problems in its introduction. The US census had intended to use the handheld computers but cost escalated and this was abandoned, with the contract being sold to Brazil. Online response is a good idea but one of the functions of census is to make sure everyone is counted accurately. A system which allowed people to enter their address without verification would be open to abuse. Therefore, households have to be verified on the ground, typically by an enumerator visit or post out. Paper forms are still necessary for those without access to Internet connections. It is also possible that the hidden nature of an administrative census means that users are not engaged with the importance of contributing their data to official statistics.

Alternatively, population estimations may be carried out remotely with GIS and remote sensing technologies.

Census and Development

According to UNFPA, "The information generated by a population and housing census – numbers of people, their distribution, their living conditions and other key data – is critical for development." This is because this type of data is essential for policymakers so that they know where to invest. Unfortunately, many countries have outdated or inaccurate data about their populations and therefore, without accurate data are unable to address the needs of their population.

UNFPA stated that,

"The unique advantage of the census is that it represents the entire statistical universe, down to the smallest geographical units, of a country or region. Planners need this information for all kinds of development work, including: assessing demographic trends; analysing socio-economic conditions; designing evidence-based poverty-reduction strategies; monitoring and evaluating the effectiveness of policies; and tracking progress toward national and internationally agreed development goals."

In addition to making policymakers aware about population issues, it is also an important tool for identifying forms of social, demographic or economic exclusions, such as inequalities relating to race, ethics and religion as well as disadvantaged groups such as those with disabilities and the poor.

An accurate census can empower local communities by providing them with the necessary information to participate in local decision-making and ensuring they are represented.

Uses of Census Data

In the nineteenth century, the first censuses collected paper enumerations that had to be collated by hand so the statistical uses were very basic. The government owned the data and were able to publish statistics themselves on the state of the nation. Uses were to measure changes in the population and apportion representation. Population estimates could be compared to those of other countries.

By the beginning of the twentieth century, censuses were recording households and some indications of their employment. In some countries, census archives are released for public examination after many decades, allowing genealogists to track the ancestry of interested people. Archives provide a substantial historical record which may challenge established notions of tradition. It is also possible to understand the societal history through job titles and arrangements for the destitute and sick.

Census Data and Research

As governments assumed responsibility for schooling and welfare, large government research departments made extensive use of census data. Actuarial estimates could be made to project populations and plan for provision in local government and regions. It was also possible for central government to allocate funding on the basis of census data. Even into the mid twentieth century, census data was only directly accessible to large government departments. However, computers meant that tabulations could be used directly by university researchers, large businesses and local government offices. They could use the detail of the data to answer new questions and add to local and specialist knowledge.

Now, census data are published in a wide variety of formats to be accessible to business, all levels of governance, media, students and teachers, charities and any citizen who is interested; researchers in particular have an interest in the role of *Census Field Officers* (CFO) and their assistants. Data can be represented visually or analysed in complex statistical models, to show the difference between certain areas, or to understand the association between different personal characteristics. Census data offer a unique insight into small areas and small demographic groups which sample data would be unable to capture with precision.

Privacy

Although the census provides a useful way of obtaining statistical information about a population, such information can sometimes lead to abuses, political or otherwise, made possible by the linking of individuals' identities to anonymous census data. This consideration is particularly important when individuals' census responses are made available in microdata form, but even aggregate-level data can result in privacy breaches when dealing with small areas and/or rare subpopulations.

For instance, when reporting data from a large city, it might be appropriate to give the average income for black males aged between 50 and 60. However, doing this for a town that only has two black males in this age group would be a breach of privacy because either of those persons, knowing his own income and the reported average, could determine the other man's income.

Typically, census data are processed to obscure such individual information. Some agencies do this by intentionally introducing small statistical errors to prevent the identification of individuals in marginal populations; others swap variables for similar respondents. Whatever measures have been taken to reduce the privacy risk in census data, new technology in the form of better electronic analysis of data poses increasing challenges to the protection of sensitive individual information. This is known as statistical disclosure control.

Another possibility is to present survey results by means of statistical models in the form of a multivariate distribution mixture. The statistical information in the form of conditional distributions (histograms) can be derived interactively from the estimated mixture model without any further access to the original database. As the final product does not contain any protected microdata, the model based interactive software can be distributed without any confidentiality concerns.

Another method is simply to release no data at all, except very large scale data directly to the central government. Different release strategies between government have led to an international project (IPUMS) to co-ordinate access to microdata and corresponding metadata. Such projects also promote standardising metadata by projects such as SDMX so that best use can be made of the minimal data available.

Historical Censuses

Egypt

Censuses in Egypt first appears in the late Middle Kingdom and develops in the New Kingdom Pharaoh Amasis, according to Herodotus, require every Egyptian to declare annually to the nomarch, "whence he gained his living". Under the Ptolemies and the Romans several censuses were conducted in Egypt by governments officials

Ancient Greece

There are several accounts of ancient Greek city states carrying out censuses.

Ancient Israel

Censuses are mentioned in the Bible. God commands a per capita tax to be paid with the census in Exodus 30:11-16 for the upkeep of the Tabernacle. The Book of Numbers is named after the counting of the Israelite population (in Numbers 1-4) according to the house of the Fathers after the exodus from Egypt. A second census was taken while the Israelite were camped in the plains of Moab, in Numbers 26.

King David performed a census that produced disastrous results (in 2 Samuel 24 and 1 Chronicles 21). His son, King Solomon, had all of the foreigners in Israel counted in 2 Chronicles 2:17.

When the Romans took over Judea in AD 6, the legate Publius Sulpicius Quirinius organised a census for tax purposes. The Gospel of Luke links the birth of Jesus to this event. Luke 2.

China

One of the world's earliest preserved censuses was held in China in AD 2 during the Han Dynasty, and is still considered by scholars to be quite accurate. Another census was held in AD 144.

India

The oldest recorded census in India is thought to have occurred around 300 BC during the reign of the Emperor Chandragupta Maurya under the leadership of Kautilya or Chanakya and Ashoka.

Rome

The word "census" originated in ancient Rome from the Latin word *censere* ("to estimate"). The census played a crucial role in the administration of the Roman Empire, as it was used to determine taxes. With few interruptions, it was usually carried out every five years. It provided a register of citizens and their property from which their duties and privileges could be listed. It is said to have been instituted by the Roman king Servius Tullius in the 6th century BC, at which time the number of arms-bearing citizens was supposedly counted at around 80,000. The 6 BC "census of Quirinius" undertaken following the imposition of direct Roman rule in Judea was partially responsible for the development of the Zealot movement and several failed rebellions against Rome that ended in the Diaspora. The 15-year indiction cycle established by Diocletian in AD 297 was based on quindecennial censuses and formed the basis for dating in late antiquity and under the Byzantine Empire.

Rashidun and Umayyad Caliphates

In the Middle Ages, the Caliphate began conducting regular censuses soon after its formation, beginning with the one ordered by the second Rashidun caliph, Umar.

Medieval Europe

The Domesday Book was undertaken in AD 1086 by William I of England so that he could properly tax the land he had recently conquered in medieval Europe. In 1183, a census was taken of the crusader Kingdom of Jerusalem, to ascertain the number of men and amount of money that could possibly be raised against an invasion by Saladin, sultan of Egypt and Syria.

Inca Empire

In the 15th century, the Inca Empire had a unique way to record census information. The Incas did not have any written language but recorded information collected during censuses and other numeric information as well as non-numeric data on quipus, strings from llama or alpaca hair or cotton cords with numeric and other values encoded by knots in a base-10 positional system.

Spanish Empire

On May 25, 1577, King Philip II of Spain ordered by royal cédula the preparation of a general description of Spain's holdings in the Indies. Instructions and a questionnaire, issued in 1577 by the Office of the Cronista Mayor, were distributed to local officials in the Viceroyalties of New Spain and Peru to direct the gathering of information. The questionnaire, composed of fifty items, was designed to elicit basic information about the nature of the land and the life of its peoples. The replies, known as "relaciones geográficas," were written between 1579 and 1585 and were returned to the Cronista Mayor in Spain by the Council of the Indies.

World Population Estimates

TABLE 3.

PRESENT POPULATION OF THE EARTH AND THE CONTINENTS
(In Millions)

Continent	According to: International Statistical Institute 1929	According to: League of Nations 1929
Asia	954	918
Europe	478	520
North America	162	161
Africa	140	146
South America	77	79
Australasia	9	9
Total	1,820	1,833

League of Nations and International Statistical Institute estimates of the world population in 1929.

The earliest estimate of the world population was made by Giovanni Battista Riccioli in 1661; the next by Johann Peter Süssmilch in 1741, revised in 1762; the third by Karl Friedrich Wilhelm Dieterici in 1859.

In 1931 Walter Willcox published a table in his book, *International Migrations: Volume II Interpretations*, that estimated the 1929 world population to be roughly 1.8 billion.

Complete count of population is called census. The observations on all the sampling units in the population are collected in a census. For example, in India, the census is conducted at every tenth year in which observations on all the persons staying in India is collected.

Sample

One or more sampling units are selected from the population according to some specified procedure. A sample consists only of a portion of the population units.

In the context of sample surveys, a collection of units like households, people, cities, countries etc. is called a finite population.

A census is a 100% sample and it is a complete count of the population.

Representative Sample

All salient features of population are present in the sample.

It goes without saying that every sample is considered as a representative sample.

For example, if a population has 30% males and 70% females, then we also expect the sample to have nearly 30% males and 70% females.

In another example, if we take out a handful of wheat from a 100 kg. bag of wheat, we expect the same quality of wheat in hand as inside the bag. Similarly, it is expected that a drop of blood will give the same information as all the blood in the body.

Sampling Frame

In statistics, a sampling frame is the source material or device from which a sample is drawn. It is a list of all those within a population who can be sampled, and may include individuals, households or institutions.

Importance of the sampling frame is stressed by Jessen and Salant and Dillman.

In many practical situations the frame is a matter of choice to the survey planner, and sometimes a critical one. [...] Some very worthwhile investigations are not undertaken at all because of the lack of an apparent frame; others, because of faulty frames, have ended in a disaster or in cloud of doubt.

—*Raymond James Jessen*

The white pages, a collection of telephone directories in the United States, was commonly used as a sampling frame for opinion polls.

Obtaining and Organizing a Sampling Frame

In the most straightforward cases, such as when dealing with a batch of material from a production run, or using a census, it is possible to identify and measure every single item in the population and to include any one of them in our sample; this is known as *direct element sampling*. However, in many other cases this is not possible; either because it is cost-prohibitive (reaching every citizen of a country) or impossible (reaching all humans alive).

Having established the frame, there are a number of ways for organizing it to improve efficiency and effectiveness. It's at this stage that the researcher should decide whether the sample is in fact to be the whole population and would therefore be a census.

This list should also facilitate access to the selected sampling units. A frame may also provide additional 'auxiliary information' about its elements; when this information is related to variables or groups of interest, it may be used to improve survey design. While not necessary for simple sampling, a sampling frame used for more advanced sample techniques, such as stratified sampling, may contain additional information (such as demographic information). For instance, an electoral register might include name and sex; this information can be used to ensure that a sample taken from that frame covers all demographic categories of interest. (Sometimes the auxiliary information is less explicit; for instance, a telephone number may provide some information about location).

Sampling Frame Qualities

An ideal sampling frame will have the following qualities:

- all units have a logical, numerical identifier
- all units can be found – their contact information, map location or other relevant information is present
- the frame is organized in a logical, systematic fashion
- the frame has additional information about the units that allow the use of more advanced sampling frames
- every element of the population of interest is present in the frame
- every element of the population is present *only once* in the frame
- no elements from outside the population of interest are present in the frame
- the data is 'up-to-date'

Types of Sampling Frames

The most straightforward type of frame is a list of elements of the population (preferably the entire population) with appropriate contact information. For example, in an opinion poll, possible sampling frames include an electoral register or a telephone directory. Other sampling frames can include employment records, school class lists, patient files in a hospital, organizations listed in a thematic database, and so on. On a more practical levels, sampling frames have the form of computer files.

Not all frames explicitly list population elements; some list only 'clusters'. For example, a street map can be used as a frame for a door-to-door survey; although it doesn't show individual houses, we can select streets from the map and then select houses on those streets. This offers some advantages: such a frame would include people who have recently moved and are not yet on the list frames discussed above, and it may be easier to use because it doesn't require storing data for every unit in the population, only for a smaller number of clusters.

Sampling Frames Problems

The sampling frame must be representative of the population and this is a question outside the scope of statistical theory demanding the judgment of experts in the particular subject matter being studied. All the above frames omit some people who will vote at the next election and contain some people who will not; some frames will contain multiple records for the same person. People not in the frame have no prospect of being sampled.

Because a cluster-based frame contains less information about the population, it may place constraints on the sample design, possibly requiring the use of less efficient sampling methods and/or making it harder to interpret the resulting data.

Statistical theory tells us about the uncertainties in extrapolating from a sample to the frame. It should be expected that sample frames, will always contain some mistakes. In some cases, this may lead to sampling bias. Such bias should be minimized, and identified, although avoiding it completely in a real world is nearly impossible. One should also not assume that sources which claim to be unbiased and representative are such.

In defining the frame, practical, economic, ethical, and technical issues need to be addressed. The need to obtain timely results may prevent extending the frame far into the future. The difficulties can be extreme when the population and frame are disjoint. This is a particular problem in forecasting where inferences about the future are made from historical data. In fact, in 1703, when Jacob Bernoulli proposed to Gottfried Leibniz the possibility of using historical mortality data to predict the probability of early death of a living man, Gottfried Leibniz recognized the problem in replying:

> Nature has established patterns originating in the return of events but only for the most part. New illnesses flood the human race, so that no matter how many experiments you have done on corpses, you have not thereby imposed a limit on the nature of events so that in the future they could not vary.
>
> —*Gottfried Leibniz*

Leslie Kish posited four basic problems of sampling frames:

1. Missing elements: Some members of the population are not included in the frame.
2. Foreign elements: The non-members of the population are included in the frame.
3. Duplicate entries: A member of the population is surveyed more than once.
4. Groups or clusters: The frame lists clusters instead of individuals.

Problems like those listed can be identified by the use of pre-survey tests and pilot studies.

List of all the units of the population to be surveyed constitutes the sampling frame. All the sampling units in the sampling frame have identification particulars. For example, all the students in a particular university listed along with their roll numbers constitute the sampling frame. Similarly, the list of households with the name of head of family or house address constitutes the sampling frame. In another example, the residents of a city area may be listed in more than one frame - as per automobile registration as well as the listing in the telephone directory.

Ways to Ensure Representativeness

There are two possible ways to ensure that the selected sample is representative.

1. Random sample or probability sample:

The selection of units in the sample from a population is governed by the laws of chance or probability. The probability of selection of a unit can be equal as well as unequal.

2. Non-random sample or purposive sample:

The selection of units in the sample from population is not governed by the probability laws.

For example, the units are selected on the basis of personal judgment of the surveyor. The persons volunteering to take some medical test or to drink a new type of coffee also constitute the sample on non- random laws.

Another type of sampling is Quota Sampling. The survey in this case is continued until a predetermined number of units with the characteristic under study are picked up.

For example, in order to conduct an experiment for rare type of disease, the survey is continued till thc required number of patients with disease are collected.

Advantages of Sampling Over Complete Enumeration

Greater accuracy

The persons involved in the collection of data are trained personals. They can collect the data more accurately if they have to collect smaller number of units than large number of unites in a given time.

Urgent information required

The data from a sample can be quickly summarized.

For example, the forecasting of the crop production can be done quickly on the basis of a sample of data than collecting first all the observations.

Feasibility

Conducting the experiment on smaller number of units, particularly when the units are destroyed, is more feasible.

For example, in determining the life of bulbs, it is more feasible to fuse minimum number of bulbs. Similarly, in any medical experiment, it is more feasible to use less number of animals.

Reduced cost and enlarged scope

Sampling involves the collection of data on smaller number of units in comparison to complete enumeration, so the cost involved in the collection of information is reduced. Further, additional information can be obtained at little cost in comparison to conducting another survey. For example, when an interviewer is collecting information on health conditions, then he/she can also ask some questions on health practices. This will provide additional information on health practices and the cost involved will be much less than conducting an entirely new survey on health practices.

Organization of work

It is easier to manage the organization of collection of smaller number of units than all the units in a census. For example, in order to draw a representative sample from a state, it is easier to manage to draw small samples from every city than drawing the sample from the whole state at a time. This ultimately results in more accuracy in the statistical inferences because better organization provides better data and in turn, improved statistical inferences are obtained.

Survey Methodology

A field of applied statistics of human research surveys, survey methodology studies the sampling of individual units from a population and the associated survey data collection techniques, such as questionnaire construction and methods for improving the number and accuracy of responses to surveys. Survey methodology includes instruments or procedures that ask one or more questions that may, or may not, be answered.

Statistical surveys are undertaken with a view towards making statistical inferences about the population being studied, and this depends strongly on the survey questions used. Polls about public opinion, public health surveys, market research surveys, government surveys and censuses are all examples of quantitative research that use contemporary survey methodology to answer questions about a population. Although censuses do not include a "sample," they do include other aspects of survey methodology, like questionnaires, interviewers, and nonresponse follow-up techniques. Surveys provide important information for all kinds of public information and research fields, e.g., marketing research, psychology, health professionals and sociology.

A single survey is made of at least a sample (or full population in the case of a census), a method of data collection (e.g., a questionnaire) and individual questions or items that become data that can be analyzed statistically. A single survey may focus on different types of topics such as preferences (e.g., for a presidential candidate), opinions (e.g., should abortion be legal?), behavior (smoking and alcohol use), or factual information (e.g., income), depending on its purpose. Since survey research is almost always based on a sample of the population, the success of the research is dependent on the representativeness of the sample with respect to a target population of interest to the researcher.

That target population can range from the general population of a given country to specific groups of people within that country, to a membership list of a professional organization, or list of students enrolled in a school system. The persons replying to a survey are called respondents, and depending on the questions asked their answers may represent themselves as individuals, their households, employers, or other organization they represent.

Survey methodology as a scientific field seeks to identify principles about the sample design, data collection instruments, statistical adjustment of data, and data processing, and final data analysis that can create systematic and random survey errors. Survey errors are sometimes analyzed in connection with survey cost. Cost constraints are sometimes framed as improving quality within cost constraints, or alternatively, reducing costs for a fixed level of quality. Survey methodology is both a scientific field and a profession, meaning that some professionals in the field focus on survey errors empirically and others design surveys to reduce them. For survey designers, the task involves making a large set of decisions about thousands of individual features of a survey in order to improve it.

The most important methodological challenges of a survey methodologist include making decisions on how to:

- Identify and select potential sample members.
- Contact sampled individuals and collect data from those who are hard to reach (or reluctant to respond).
- Evaluate and test questions.
- Select the mode for posing questions and collecting responses.
- Train and supervise interviewers (if they are involved).
- Check data files for accuracy and internal consistency.
- Adjust survey estimates to correct for identified errors.

Selecting Samples

The sample is chosen from the sampling frame, which consists of a list of all members of the population of interest. The goal of a survey is not to describe the sample, but the larger population. This generalizing ability is dependent on the representativeness of the sample, as stated above. Each member of the population is termed an element. There are frequent difficulties one encounters while choosing a representative sample. One common error that results is selection bias. Selection bias results when the procedures used to select a sample result in over representation or under representation of some significant aspect of the population. For instance, if the population of interest consists of 75% females, and 25% males, and the sample consists of 40% females and 60% males, females are under represented while males are overrepresented. In order to minimize selection biases, stratified random sampling is often used. This is when the population is divided into sub-populations called strata, and random samples are drawn from each of the strata, or elements are drawn for the sample on a proportional basis.

Modes of Data Collection

There are several ways of administering a survey. The choice between administration modes is influenced by several factors, including

1. costs,
2. coverage of the target population,
3. flexibility of asking questions,
4. respondents' willingness to participate,
5. response accuracy.

Different methods create mode effects that change how respondents answer, and different methods have different advantages. The most common modes of administration can be summarized as:

- Telephone
- Mail (post)
- Online surveys
- Personal in-home surveys
- Personal mall or street intercept survey
- Hybrids of the above.

Research Designs

There are several different designs, or overall structures, that can be used in survey research. The three general types are cross-sectional, successive independent samples, and longitudinal studies.

Cross-sectional Studies

In cross-sectional studies, a sample (or samples) is drawn from the relevant population and studied once. A cross-sectional study describes characteristics of that population at one time, but cannot give any insight as to the causes of population characteristics because it is a predictive, correlational design.

Successive Independent Samples Studies

A successive independent samples design draws multiple random samples from a population at one or more times. This design can study changes within a population, but not changes within individuals because the same individuals are not surveyed more than once. Such studies cannot, therefore, identify the causes of change over time necessarily. For successive independent samples designs to be effective, the samples must be drawn from the same population, and must be equally representative of it. If the samples are not comparable, the changes between samples may be due to demographic characteristics rather than time. In addition, the questions must be asked in the same way so that responses can be compared directly.

Longitudinal Studies

Longitudinal studies take measure of the same random sample at multiple time points. Unlike with a successive independent samples design, this design measures the differences in individual participants' responses over time. This means that a researcher can potentially assess the reasons for response changes by assessing the differences in respondents' experiences. Longitudinal studies are the easiest way to assess the effect of a naturally occurring event, such as divorce that cannot be tested experimentally. However, longitudinal studies are both expensive and difficult to do. It's harder to find a sample that will commit to a months- or years-long study than a 15-minute interview, and participants frequently leave the study before the final assessment. This attrition of participants is not random, so samples can become less representative with successive assessments. To account for this, a researcher can compare the respondents who left the survey to those that did not, to see if they are statistically different populations. Respondents may also try to be self-consistent in spite of changes to survey answers.

Questionnaires

Questionnaires are the most commonly used tool in survey research. However, the results of a particular survey are worthless if the questionnaire is written inadequately. Questionnaires should produce valid and reliable demographic variable measures and should yield valid and reliable individual disparities that self-report scales generate.

Questionnaires as Tools

A variable category that is often measured in survey research are demographic variables, which are used to depict the characteristics of the people surveyed in the sample. Demographic variables include such measures as ethnicity, socioeconomic status, race, and age. Surveys often assess the preferences and attitudes of individuals, and many employ self-report scales to measure people's opinions and judgements about different items presented on a scale. Self-report scales are also used to examine the disparities among people on scale items. These self-report scales, which are usually presented in questionnaire form, are one of the most used instruments in psychology, and thus it is important that the measures be constructed carefully, while also being reliable and valid.

Reliability and Validity of Self-report Measures

Reliable measures of self-report are defined by their consistency. Thus, a reliable self-report measure produces consistent results every time it is executed. A test's reliability can be measured a few ways. First, one can calculate a test-retest reliability. A test-retest reliability entails conducting the same questionnaire to a large sample at two different times. For the questionnaire to be considered reliable, people in the sample do not have to score identically on each test, but rather their position in the score distribution should be similar for both the test and the retest. Self-report measures will generally be more reliable when they have many items measuring a construct. Furthermore, measurements will be more reliable when the factor being measured has greater variability among the individuals in the sample that are being tested. Finally, there will be greater reliability when instructions for the completion of the questionnaire are clear and when there are limited distractions in the testing environment. Contrastingly, a questionnaire is valid if what it measures is what it had originally planned to measure. Construct validity of a measure is the degree to which it measures the theoretical construct that it was originally supposed to measure.

Composing a Questionnaire

Six steps can be employed to construct a questionnaire that will produce reliable and valid results. First, one must decide what kind of information should be collected. Second, one must decide how to conduct the questionnaire. Thirdly, one must construct a first draft of the questionnaire. Fourth, the questionnaire should be revised. Next, the questionnaire should be pretested. Finally, the questionnaire should be edited and the procedures for its use should be specified.

Guidelines for the Effective Wording of Questions

The way that a question is phrased can have a large impact on how a research participant will answer the question. Thus, survey researchers must be conscious of their wording when writing survey questions. It is important for researchers to keep in mind that different individuals, cultures, and subcultures can interpret certain words and phrases differently from one another. There are two different types of questions that survey researchers use when writing a questionnaire: free response questions and closed questions. Free response questions are open-ended, whereas closed questions are usually multiple choice. Free response questions are beneficial because they allow the responder greater flexibility, but they are also very difficult to record and score, requiring extensive coding. Contrastingly, closed questions can be scored and coded much easier, but they diminish expressivity and spontaneity of the responder. In general, the vocabulary of the questions should be very simple and direct, and most should be less than twenty words. Each question should be edited for "readability" and should avoid leading or loaded questions. Finally, if multiple items are being used to measure one construct, the wording of some of the items should be worded in the opposite direction to evade response bias.

A respondent's answer to an open-ended question can be coded into a response scale afterwards, or analysed using more qualitative methods.

Order of Questions

Survey researchers should carefully construct the order of questions in a questionnaire. For questionnaires that are self-administered, the most interesting questions should be at the beginning of the questionnaire to catch the respondent's attention, while demographic questions should be near the end. Contrastingly, if a survey is being administered over the telephone or in person, demographic questions should be administered at the beginning of the interview to boost the respondent's confidence. Another reason to be mindful of question order may cause a survey response effect in which one question may affect how people respond to subsequent questions as a result of priming.

Nonresponse Reduction

The following ways have been recommended for reducing nonresponse in telephone and face-to-face surveys:

- Advance letter. A short letter is sent in advance to inform the sampled respondents about the upcoming survey. The style of the letter should be personalized but not overdone. First, it announces that a phone call will be made/ or an interviewer wants to make an appointment to do the survey face-to-face. Second, the research topic will be described. Last, it allows both an expression of the surveyor's appreciation of cooperation and an opening to ask questions on the survey.
- Training. The interviewers are thoroughly trained in how to ask respondents questions, how to work with computers and making schedules for callbacks to respondents who were not reached.
- Short introduction. The interviewer should always start with a short introduction about him or herself. She/he should give her name, the institute she is working for, the length of the interview and goal of the interview. Also it can be useful to make clear that you are not selling anything: this has been shown to lead to a slightly higher responding rate.

- Respondent-friendly survey questionnaire. The questions asked must be clear, non-offensive and easy to respond to for the subjects under study.

Brevity is also often cited as increasing response rate. A 1996 literature review found mixed evidence to support this claim for both written and verbal surveys, concluding that other factors may often be more important. A 2010 study looking at 100,000 on-line surveys found response rate dropped by about 3% at 10 questions and about 6% at 20 questions, with drop-off slowing (for example, only 10% reduction at 40 questions). Other studies showed that quality of response degraded toward the end of long surveys.

Interviewer Effects

Survey methodologists have devoted much effort to determining the extent to which interviewee responses are affected by physical characteristics of the interviewer. Main interviewer traits that have been demonstrated to influence survey responses are race, gender, and relative body weight (BMI). These interviewer effects are particularly operant when questions are related to the interviewer trait. Hence, race of interviewer has been shown to affect responses to measures regarding racial attitudes, interviewer sex responses to questions involving gender issues, and interviewer BMI answers to eating and dieting-related questions. While interviewer effects have been investigated mainly for face-to-face surveys, they have also been shown to exist for interview modes with no visual contact, such as telephone surveys and in video-enhanced web surveys. The explanation typically provided for interviewer effects is social desirability bias: survey participants may attempt to project a positive self-image in an effort to conform to the norms they attribute to the interviewer asking questions. Interviewer effects are one example survey response effects.

Type of Surveys

1. Demographic surveys

These surveys are conducted to collect the demographic data, e.g., household surveys, family size, number of males in families, etc.

Such surveys are useful in the policy formulation for any city, state or country for the welfare of the people.

2. Educational surveys

These surveys are conducted to collect the educational data, e.g., how many children go to school, how many persons are graduate, etc.

Such surveys are conducted to examine the educational programs in schools and colleges. Generally, schools are selected first and then the students from each school constitue the sample.

3. Economic surveys

These surveys are conducted to collect the economic data, e.g., data related to export and import of goods, industrial production, consumer expenditure etc. Such data is helpful in constructing the indices indicating the growth in a particular sector of economy or even the overall economic growth of the country.

4. Employment surveys

These surveys are conducted to collect the employment related data, e.g., employment rate, labour conditions, wages, etc. in a city, state or country. Such data helps in constructing various indices to know the employment conditions among the people.

5. Health and nutrition surveys

These surveys are conducted to collect the data related to health and nutrition issues, e.g., number of visits to doctors, food given to children, nutritional value etc. Such surveys are conducted in cities, states as well as countries by national and international organizations like UNICEF, WHO etc.

6. Agricultural surveys

These surveys are conducted to collect the agriculture related data to estimate, e.g., the acreage and production of crops, livestock numbers, use of fertilizers, use of pesticides and other related topics. The government bases its planning related to the food issues for the people based on such surveys.

7. Marketing surveys

These surveys are conducted to collect the data related to marketing. They are conducted by major companies, manufacturers or those who provide services to consumer etc. Such data is used for knowing the satisfaction and opinion of consumers as well as in developing the sales, purchase and promotional activities etc.

8. Election surveys

These surveys are conducted to study the outcome of an election or a poll. For example, such polls are conducted in democratic countries to have the opinions of people about any candidate who is contesting the election.

9. Public polls and surveys

These surveys are conducted to collect the public opinion on any particular issue. Such surveys are generally conducted by news media and agencies which conduct polls and surveys on current topics of interest to public.

10. Campus surveys

These surveys are conducted on the students of any educational institution to study about the educational programs, living facilities, dining facilities, sports activities, etc.

Principal Steps in a Sample Survey

The broad steps to conduct any sample survey are as follows:

Objective of the Survey

The objective of the survey has to be clearly defined and well understood by the person planning to conduct it. It is expected from the statistician to be well versed with the issues to be addressed in consultation with the person who wants to get the survey conducted. In complex surveys, sometimes the objective is forgotten and data is collected on those issues which are far away from the objectives.

Population to be Sampled

Based on the objectives of the survey, decide the population from which the information can be obtained. For example, population of farmers is to be sampled for an agricultural survey whereas the population of patients has to be sampled for determining the medical facilities in a hospital.

Data to be Collected

It is important to decide that which data is relevant for fulfilling the objectives of the survey and to note that no essential data is omitted. Sometimes, too many questions are asked and some of their outcomes are never utilized. This lowers the quality of the responses and in turn results in lower efficiency in statistical inferences.

Degree of Precision Required

The results of any sample survey are always subjected to some uncertainty. Such uncertainty can be reduced by taking larger samples or using superior instruments. This involves more cost and more time. So, it is very important to decide about the required degree of precision in data. This needs to be conveyed to the surveyor also.

Method of Measurement

The choice of measuring instrument and method to measure the data from the population needs to be specified clearly. For example, the data has to be collected through interview, questionnaire, personal visit, combination of any of these approaches, etc. The forms in which the data is to be recorded so that the data can be transferred to mechanical equipment for easily creating the data summary etc. is also needed to be prepared accordingly.

The Frame

The sampling frame has to be clearly specified. The population is divided into sampling units such that the units cover the whole population and every sampling unit is tagged with identification. The list of all sampling units is called the frame. The frame must cover the whole population and the units must not overlap each other in the sense that every element in the population must belong to one and only one unit. For example, the sampling unit can be an individual member in the family or the whole family.

Selection of Sample

The size of the sample needs to be specified for the given sampling plan. This helps in determining and comparing the relative cost and time of different sampling plans. The method and plan adopted for drawing a representative sample should also be detailed.

The Pre-test

It is advised to try the questionnaire and field methods on a small scale. This may reveal some troubles and problems beforehand which the surveyor may face in field in large scale surveys.

Organization of the Field Work

How to conduct the survey, how to handle business administrative issues, providing proper training to surveyors, procedures, plans for handling the nonresponse and missing observations etc. are some of the issues which need to be addressed for organizing the survey work in the fields. The procedure for early checking of the quality of return should be prescribed. It should be clarified how to handle the situation when the respondent is not available.

Analysis of Data

It is to be noted that based on the objectives of the data, the suitable statistical tool is decided which can answer the relevant questions. In order to use the statistical tool, a valid data set is required and this dictates the choice of responses to be obtained for the questions in the questionnaire, e.g., the data has to be qualitative, quantitative, nominal, ordinal etc. After getting the completed questionnaire back, it needs to be edited to amend the recording errors and delete the erroneous data. The tabulating procedures, methods of estimation and tolerable amount of error in the estimation needs to be decided before the start of survey. Different methods of estimation may be available to get the answer of the same query from the same data set. So the data needs to be collected which is compatible with the chosen estimation procedure.

Information Gained for Future Surveys

The completed surveys work as guide for improved sample surveys in future. Beside this they also supply various types of prior information required to use various statistical tools, e.g., mean, variance, nature of variability, cost involved etc. Any completed sample survey acts as a potential guide for the surveys to be conducted in the future. It is generally seen that the things always do not go in the same way in any complex survey as planned earlier. Such precautions and alerts help in avoiding the mistakes in the execution of future surveys.

Variability Control in Sample Surveys

The variability control is an important issue in any statistical analysis. A general objective is to draw statistical inferences with minimum variability. There are various types of sampling schemes which are adopted in different conditions. These schemes help in controlling the variability at different stages. Such sampling schemes can be classified in the following way.

Before Selection of Sampling Units

- Stratified sampling
- Cluster sampling
- Two stage sampling
- Double sampling etc.

At the Time of Selection of Sampling Units

- Systematic sampling
- Varying probability sampling

After the Selection of Sampling Units

- Ratio method of estimation
- Regression method of estimation

Methods of Data Collection

Physical Observations and Measurements:

The surveyor contacts the respondent personally through meeting. He observes the sampling unit and records the data. The surveyor can always use his prior experience to collect the data in a better way. For example, a young man telling his age as 60 years can easily be observed and corrected by the surveyor.

Personal Interview:

The surveyor is supplied with a well prepared questionnaire. The surveyor goes to the respondents and asks the same questions mentioned in the questionnaire. The data in the questionnaire is then filled up accordingly based on the responses from the respondents.

Mail Enquiry:

The well prepared questionnaire is sent to the respondents through postal mail, e-mail, etc. The respondents are requested to fill up the questionnaires and send it back. In case of postal mail, many times the questionnaires are accompanied by a self addressed envelope with postage stamps to avoid any non-response due to the cost of postage.

Web Based Enquiry:

The survey is conducted online through internet based web pages. There are various websites which provide such facility. The questionnaires are to be in their formats and the link is sent to the respondents through email. By clicking on the link, the respondent is brought to the concerned website and the answers are to be given online. These answers are recorded and responses as well as their statistics is sent to the surveyor. The respondents should have internet connection to support the data collection with this procedure.

Registration:

The respondent is required to register the data at some designated place. For example, the number of births and deaths along with the details provided by the family members are recorded at city municipal office which are provided by the family members.

Transcription From Records:

The sample of data is collected from the already recorded information. For example, the details of the number of persons in different families or number of births/ deaths in a city can be obtained from the city municipal office directly. The methods in (1) to (5) provide primary data which means collecting the data directly from the source. The method in (6) provides the secondary data which means getting the data from the primary sources.

Statistical Unit

A unit in a statistical analysis refers to one member of a set of entities being studied. It is the material source for the mathematical abstraction of a "random variable". Common examples of a unit would be a single person, animal, plant, manufactured item, or country that belongs to a larger collection of such entities being studied.

Units are often referred to as being either experimental units, sampling units or, more generally, units of observation:

- An "experimental unit" is typically thought of as one member of a set of objects that are initially equivalent, with each object then subjected to one of several experimental treatments.
- A "sampling unit" is typically thought of as an object that has been sampled from a statistical population. This term is commonly used in opinion polling and survey sampling.

In most statistical studies, the goal is to generalize from the observed units to a larger set consisting of all comparable units that exist but are not directly observed. For example, if we randomly sample 100 people and ask them which candidate they intend to vote for in an election, our main interest is in the voting behavior of all eligible voters, not exclusively on the 100 observed units.

In some cases, the observed units may not form a sample from any meaningful population, but rather constitute a convenience sample, or may represent the entire population of interest. In this situation, we may study the units descriptively, or we may study their dynamics over time. But it typically does not make sense to talk about generalizing to a larger population of such units. Studies involving countries or business firms are often of this type. Clinical trials also typically use convenience samples, however the aim is often to make inferences about the efficacy of treatments in other patients, and given the inclusion and exclusion criteria for some clinical trials, the sample may not be representative of the majority of patients with the condition or disease.

In simple data sets, the units are in one-to-one correspondence with the data values. In more complex data sets, multiple measurements are made for each unit. For example, if blood pressure measurements are made daily for a week on each subject in a study, there would be seven data values for each statistical unit. Multiple measurements taken on an individual are not independent (they will be more alike compared to measurements taken on different individuals). Ignoring these dependencies during the analysis can lead to an inflated sample size or pseudoreplication.

While a *unit* is often the lowest level at which observations are made, in some cases, a *unit* can be further decomposed as a statistical assembly.

Many statistical analyses use quantitative data that have units of measurement. This is a distinct and non-overlapping use of the term "unit."

References

- Carl-Erik Särndal; Bengt Swensson; Jan Wretman (2003). Model assisted survey sampling. Springer. pp. 9–12. ISBN 978-0-387-40620-6. Retrieved 07, April 2020
- Flores-Macias, F.; Lawson, C. (2008). "Effects of interviewer gender on survey responses: Findings from a household survey in Mexico". International Journal of Public Opinion Research. 20 (1): 100–110. doi:10.1093/ijpor/edn007

- Eisinga, R.; Te Grotenhuis, M.; Larsen, J.K.; Pelzer, B.; Van Strien, T. (2011). "BMI of interviewer effects". International Journal of Public Opinion Research. 23 (4): 530–543. doi:10.1093/ijpor/edr026
- Hill, M.E (2002). "Race of the interviewer and perception of skin color: Evidence from the multi-city study of urban inequality". American Sociological Review. 67 (1): 99–108. JSTOR 3088935. doi:10.2307/3088935
- Groves, R.M.; Fowler, F. J.; Couper, M.P.; Lepkowski, J.M.; Singer, E.; Tourangeau, R. (2009). Survey Methodology. New Jersey: John Wiley & Sons. ISBN 978-1-118-21134-2
- Roger Sapsford; Victor Jupp (29 March 2006). Data collection and analysis. SAGE. pp. 28–. ISBN 978-0-7619-4363-1. Retrieved 11, June 2020
- Shahrokh Esfahani, Mohammad; Dougherty, Edward (2014). "Effect of separate sampling on classification accuracy". Bioinformatics. 30 (2): 242–250. PMID 24257187. doi:10.1093/bioinformatics/btt662

5

Statistical Data

The branch of statistics which deals with accumulation and interpretation of data to uncover the patterns and trends is known as statistical analysis. This chapter unfolds types of data, primary data, secondary data, quantitative and qualitative data, census method and survey method. These topics are crucial for a complete understanding of the subject.

Statistical Analysis

Statistical data analysis is a procedure of performing various statistical operations. It is a kind of quantitative research, which seeks to quantify the data, and typically, applies some form of statistical analysis. Quantitative data basically involves descriptive data, such as survey data and observational data.

Statistical data analysis generally involves some form of statistical tools, which a layman cannot perform without having any statistical knowledge. There are various software packages to perform statistical data analysis. This software includes Statistical Analysis System (SAS), Statistical Package for the Social Sciences (SPSS), Stat soft, etc.

Data in statistical data analysis consists of variable(s). Sometimes the data is univariate or multivariate. Depending upon the number of variables, the researcher performs different statistical techniques.

If the data in statistical data analysis is multiple in numbers, then several multivariates can be performed. These are factor statistical data analysis, discriminant statistical data analysis, etc. Sim ilarly, if the data is singular in number, then the univariate statistical data analysis is performed. This includes t test for significance, z test, f test, ANOVA one way, etc.

The data in statistical data analysis is basically of 2 types, namely, continuous data and discreet data. The continuous data is the one that cannot be counted. For example, intensity of a light can be measured but cannot be counted. The discreet data is the one that can be counted. For example, the number of bulbs can be counted.

The continuous data in statistical data analysis is distributed under continuous distribution function, which can also be called the probability density function, or simply pdf.

The discreet data in statistical data analysis is distributed under discreet distribution function, which can also be called the probability mass function or simple pmf.

We use the word 'density' in continuous data of statistical data analysis because density cannot be counted, but can be measured. We use the word 'mass' in discreet data of statistical data analysis because mass cannot be counted.

There are various pdf's and pmf's in statistical data analysis. For example, Poisson distribution is the commonly known pmf, and normal distribution is the commonly known pdf.

These distributions in statistical data analysis help us to understand which data falls under which distribution. If the data is about the intensity of a bulb, then the data would be falling in Poisson distribution.

There is a major task in statistical data analysis, which comprises of statistical inference. The statistical inference is mainly comprised of two parts: estimation and tests of hypothesis.

Estimation in statistical data analysis mainly involves parametric data—the data that consists of parameters. On the other hand, tests of hypothesis in statistical data analysis mainly involve non parametric data— the data that consists of no parameters.

Types of Data

Categorical Data

Categorical data represents characteristics. Therefore it can represent things like a person's gender, language etc. Categorical data can also take on numerical values (Example: 1 for female and 0 for male). Note that those numbers don't have mathematical meaning.

Nominal Data

Nominal values represent discrete units and are used to label variables, that have no quantitative value. Just think of them as "labels". Note that nominal data that has no order. Therefore if you would change the order of its values, the meaning would not change. You can see two examples of nominal features below:

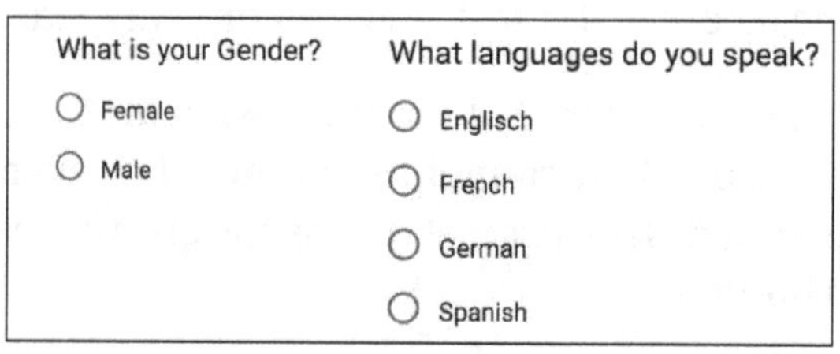

The left feature that describes a persons gender would be called „dichotomous", which is a type of nominal scales that contains only two categories.

Ordinal Data

Ordinal values represent discrete and ordered units. It is therefore nearly the same as nominal data, except that it's ordering matters. You can see an example below:

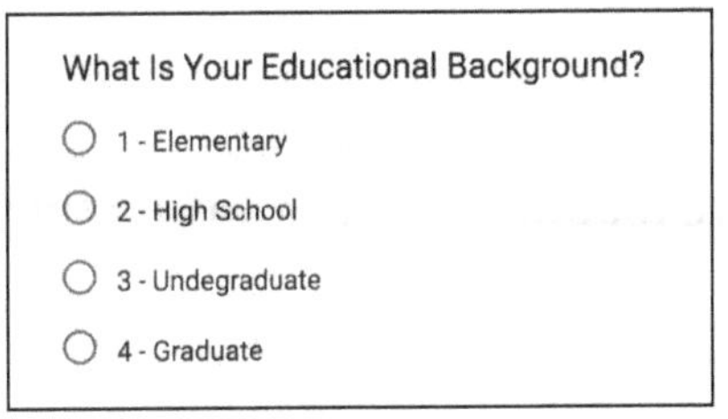

What Is Your Educational Background?

- 1 - Elementary
- 2 - High School
- 3 - Undegraduate
- 4 - Graduate

Note that the difference between Elementary and High School is different than the difference between High School and College. This is the main limitation of ordinal data, the differences between the values is not really known. Because of that, ordinal scales are usually used to measure non-numeric features like happiness, customer satisfaction and so on.

Numerical Data

1. Discrete Data

We speak of discrete data if its values are distinct and separate. In other words: We speak of discrete data if the data can only take on certain values. This type of data can't be measured but it can be counted. It basically represents information that can be categorized into a classification. An example is the number of heads in 100 coin flips.

You can check by asking the following two questions whether you are dealing with discrete data or not: Can you count it and can it be divided up into smaller and smaller parts?

2. Continuous Data

Continuous Data represents measurements and therefore their values can't be counted but they can be measured. An example would be the height of a person, which you can describe by using intervals on the real number line.

3. Interval Data

Interval values represent ordered units that have the same difference. Therefore we speak of interval data when we have a variable that contains numeric values that are ordered and where we know the exact differences between the values. An example would be a feature that contains temperature of a given place like you can see below:

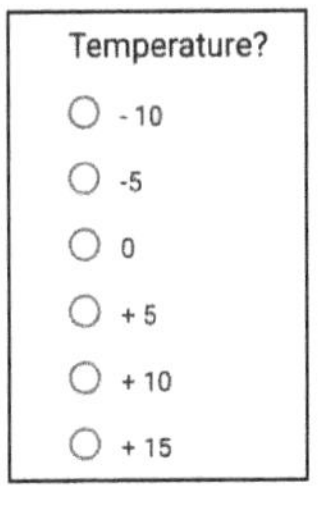

Temperature?

- -10
- -5
- 0
- +5
- +10
- +15

The problem with interval values data is that they don't have a „true zero". That means in regards to our example, that there is no such thing as no temperature. With interval data, we can add and subtract, but we cannot multiply, divide or calculate ratios. Because there is no true zero, a lot of descriptive and inferential statistics can't be applied.

Ratio Data

Ratio values are also ordered units that have the same difference. Ratio values are the same as interval values, with the difference that they do have an absolute zero. Good examples are height, weight, length etc.

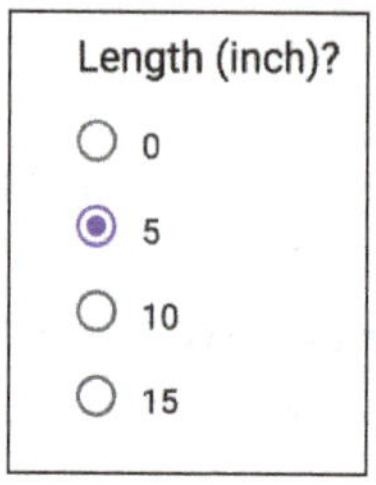

Statistical Methods

Nominal Data

When you are dealing with nominal data, you collect information through:

Frequencies: The Frequency is the rate at which something occurs over a period of time or within a dataset.

Proportion: You can easily calculate the proportion by dividing the frequency by the total number of events. (e.g how often something happened divided by how often it could happen).

Percentage

Visualization Methods: To visualize nominal data you can use a pie chart or a bar chart.

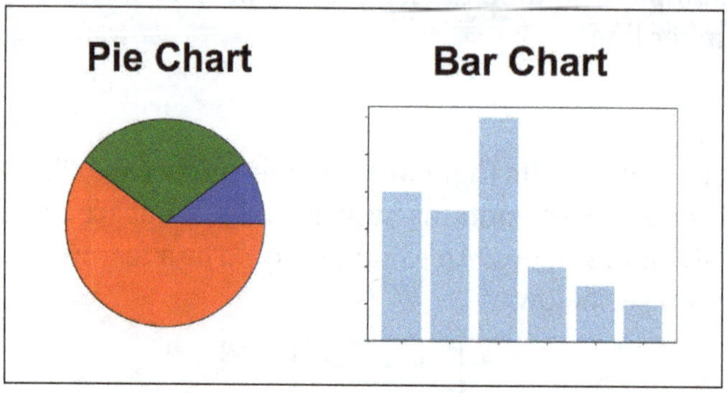

In Data Science, you can use one hot encoding, to transform nominal data into a numeric feature.

Ordinal Data

When you are dealing with ordinal data, you can use the same methods like with nominal data, but you also have access to some additional tools. Therefore you can summarize your ordinal data with frequencies, proportions, percentages. And you can visualize it with pie and bar charts. Additionally, you can use percentiles, median, mode and the interquartile range to summarize your data.

In Data Science, you can use one label encoding, to transform ordinal data into a numeric feature.

Continuous Data

When you are dealing with continuous data, you can use the most methods to describe your data. You can summarize your data using percentiles, median, interquartile range, mean, mode, standard deviation, and range.

Visualization Methods: To visualize continuous data, you can use a histogram or a box-plot. With a histogram, you can check the central tendency, variability, modality, and kurtosis of a distribution. Note that a histogram can't show you if you have any outliers. This is why we also use box-plots.

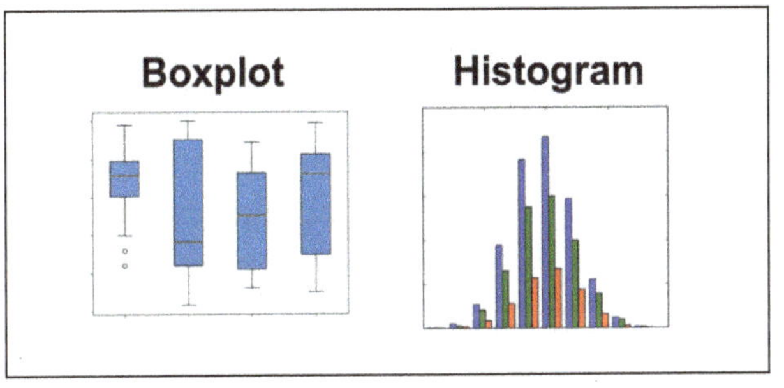

Primary Data

Primary data is data originated for the first time by the researcher through direct efforts and experience, specifically for the purpose of addressing his research problem. Also known as the first hand or raw data. Primary data collection is quite expensive, as the research is conducted by the organisation or agency itself, which requires resources like investment and manpower. The data collection is under direct control and supervision of the investigator.

Sources of Primary Data

The sources of primary data are primary units such as basic experimental units, individuals, households. Following methods are used to collect data from primary units usually and these methods depends on the nature of the primary unit. Published data and the data collected in the past is called secondary data.

- Personal Investigation: The researcher conducts the experiment or survey himself/herself and collected data from it. The collected data is generally accurate and reliable. This method of collecting primary data is feasible only in case of small scale laboratory, field experiments or pilot surveys and is not practicable for large scale experiments and surveys because it take too much time.
- Through Investigators: The trained (experienced) investigators are employed to collect the required data. In case of surveys, they contact the individuals and fill in the questionnaires after asking the required information, where a questionnaire is an inquiry form having a number of questions designed to obtain information from the respondents. This method of collecting data is usually employed by most of the organizations and its gives reasonably accurate information but it is very costly and may be time taking too.

- Through Questionnaire: The required information (data) is obtained by sending a questionnaire (printed or soft form) to the selected individuals (respondents) (by mail) who fill in the questionnaire and return it to the investigator. This method is relatively cheap as compared to "through investigator" method but non-response rate is very high as most of the respondents don't bother to fill in the questionnaire and send it back to investigator.
- Through Local Sources: The local representatives or agents are asked to send requisite information who provide the information based upon their own experience. This method is quick but it gives rough estimates only.
- Through Telephone: The information may be obtained by contacting the individuals on telephone. It's a Quick and provide accurate required information.
- Through Internet: With the introduction of information technology, the people may be contacted through internet and the individuals may be asked to provide the pertinent information. Google survey is widely used as online method for data collection now a day. There are many paid online survey services too.

Some Advantages of using Primary data:

1. The investigator collects data specific to the problem under study.
2. There is no doubt about the quality of the data collected (for the investigator).
3. If required, it may be possible to obtain additional data during the study period.

Some Disadvantages of using Primary data (for reluctant/uninterested investigators):

The investigator has to contend with all the hassles of data collection

- Deciding why, what, how, when to collect
- Getting the data collected (personally or through others)
- Getting funding and dealing with funding agencies
- Ethical considerations (consent, permissions, etc.)

Ensuring the data collected is of a high standard

- All desired data is obtained accurately, and in the format it is required in
- There is no fake/cooked up data
- Unnecessary/useless data has not been included

Cost of obtaining the data is often the major expense in studies

Secondary Data

Secondary data implies second-hand information which is already collected and recorded by any person other than the user for a purpose, not relating to the current research problem.

Sources of Secondary Data

The secondary data may be available from the following sources:

- Government Organizations: Federal and Provincial Bureau of Statistics, Crop Reporting Service-Agriculture Department, Census and Registration Organization etc.
- Semi-Government Organization: Municipal committees, District Councils, Commercial and Financial Institutions like banks etc.
- Teaching and Research Organizations.
- Research Journals and Newspapers.
- Internet.

Some Advantages of using Secondary data:

1. The data's already there- no hassles of data collection.
2. It is less expensive.
3. The investigator is not personally responsible for the quality of data.

Some disadvantages of using Secondary data:

1. The investigator cannot decide what is collected (if specific data about something is required, for instance).
2. One can only hope that the data is of good quality.
3. Obtaining additional data (or even clarification) about something is not possible (most often).

Quantitative and Qualitative Data

Qualitative Data

Qualitative data collection is a method in which the characteristics, attributes, properties, qualities, etc. of a phenomenon or thing is described. It is the description of data in a language rather than in numbers. This method does not measure the characteristics but describes them. For example; Favorite color = blue.

It is also sometimes referred to as "categorical data." It does not focus on drawing any inferences. It only deals with data that can be observed like texture, taste, smell, beauty, but is not measured.

Qualitative data, in recent years, has lost reliability to some extent and has come under criticism, but they provide a better description and thus have more validity to them. Research uses a combination of qualitative and quantitative methods because the qualitative data and description backs up the numerical data with the help of better explanations and information.

Quantitative Data

Quantitative data collection is a method in which data that can be numerically counted or expressed is collected. This data is useful for experiments, manipulated analysis, etc. and is represented by histograms, tables, charts, and graphs. It deals with measurements like height, length, volume, area, humidity, temperature, etc.

For example;

Height = 2.8m. Or sometimes they represent the exact number like,

Number of students = 234.

This type of data is associated with some type of scale measurement. The most commonly used scale for this data is a ratio scale. Another general scale measurement is the interval scale.

Quantitative data is criticized for its lack of in-depth description thus it is used by researchers along with qualitative data to back up its reliability with explanations of the qualitative information.

Importance of Quantitative and Qualitative Data

Quantitative and qualitative data provide different outcomes, and are often used together to get a full picture of a population. For example, if data are collected on annual income (quantitative), occupation data (qualitative) could also be gathered to get more detail on the average annual income for each type of occupation.

Quantitative and qualitative data can be gathered from the same data unit depending on whether the variable of interest is numerical or categorical. For example:

Data unit	Numeric variable	= Quantitative data	Categorical variable	= Qualitative data
A person	"*How many* children do you have?"	4 children	"In *which country* were your children born?"	Australia
	"*How much* do you earn?"	$60,000 p.a.	"*What* is your occupation?"	Photographer
	"How many hours do you work?"	38 hours per week	"Do you work *full-time or part-time*?"	Full-time
A house	"*How many* square metres is the house?"	200 square metres	"In *which city or town* is the house located?"	Brisbane
A business	"*How many* workers are currently employed?"	264 employees	"*What* is the *industry* of the business?"	Retail
A farm	"*How many* milk cows are located on the farm?	36 cows	"*What* is the main *activity* of the farm?"	Dairy

Uses of Quantitative and Qualitative Data

It is important to identify whether the data are quantitative or qualitative as this affects the statistics that can be produced.

Frequency counts: The number of times an observation occurs (frequency) for a data item (variable) can be shown for both quantitative and qualitative data.

The graphs below arrange the quantitative and qualitative data to show the frequency distribution of the data.

Quantitative Data

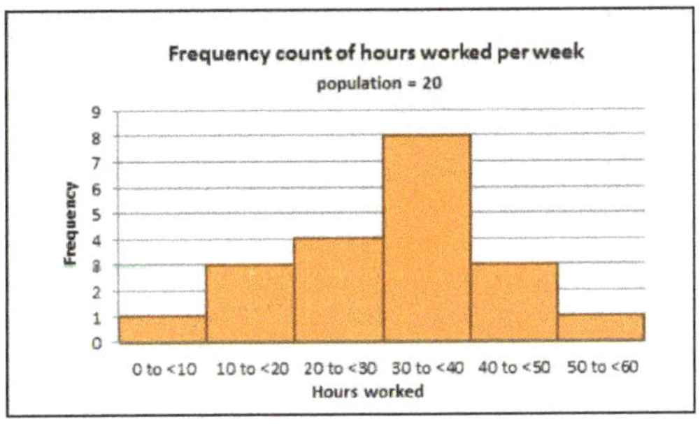

Qualitative Data

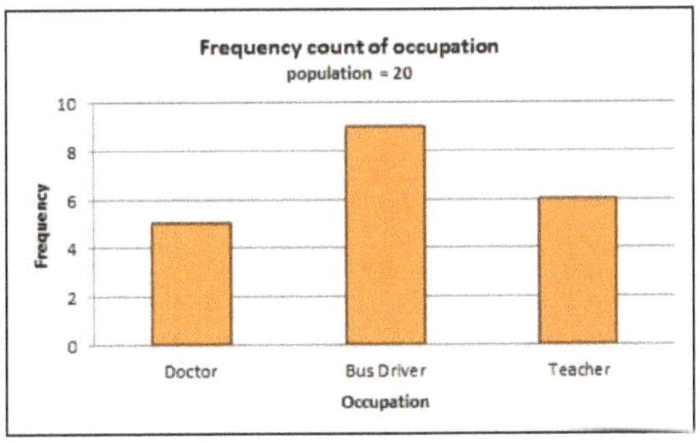

As absolute frequencies can be calculated on quantitative and qualitative data, relative frequencies can also be produced, such as percentages, proportions, rates and ratios. For example, the graphs above show 4 people (20%) worked less than 30 hours per week, and 6 people (30%) are teachers.

Descriptive (Summary) Statistics

Statistics that describe or summarise can be produced for quantitative data and to a lesser extent for qualitative data.

As quantitative data are always numeric they can be ordered, added together, and the frequency of an observation can be counted. Therefore, all descriptive statistics can be calculated using quantitative data.

As qualitative data represent individual (mutually exclusive) categories, the descriptive statistics that can be calculated are limited, as many of these techniques require numeric values which can be logically ordered from lowest to highest and which express a count.

Mode can be calculated, as it it the most frequency observed value. Median, measures of shape, measures of spread such as the range and interquartile range require an ordered data set with a logical low-end value and high-end value. Variance and standard deviation require the mean to be calculated, which is not appropriate for categorical variables as they have no numerical value.

Inferential Statistics

By making inferences about quantitative data from a sample, estimates or projections for the total population can be produced.

Quantitative data can be used to inform broader understandings of a population, or to consider how that population may change or progress into the future.

For example, a simple income projection for an employee in 2015 may be inferred from the rate of change for data collected in 2000, 2005, and 2010.

As shown in the graph below, data collected over time indicates a 5% increase every five years. Therefore, if the rate of increase continues to follow the same pattern, it can be projected that the annual income for that employee in 2015 will be $46,305; which is the 2010 wage of $44,100 increased by an additional 5%.

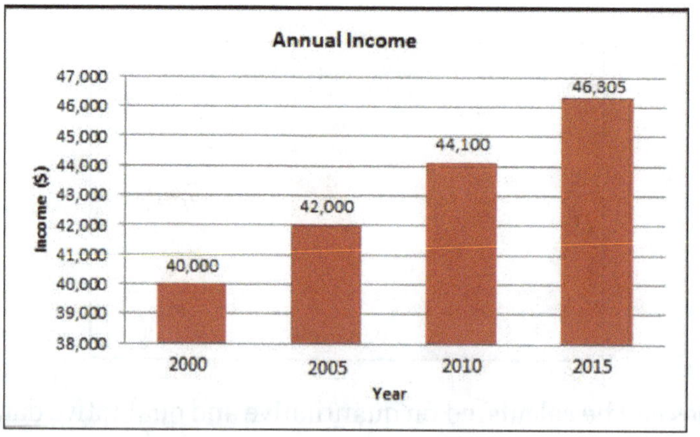

Qualitative data are not compatible with inferential statistics as all techniques are based on numeric values.

Census Method

The Census Method is also called as a Complete Enumeration Survey Method wherein each and every item in the universe is selected for the data collection. The universe might constitute a particular place, a group of people or any specific locality which is the complete set of items and which are of interest in any particular situation.

The census method is most commonly used by the government in connection with the national population, housing census, agriculture census, etc. where the vast knowledge about these fields is required. Whenever the entire population is studied to collect the detailed data about every unit, then the census method is applied.

The census method can be applied in a situation where the separate data for every unit in the population is to be collected, such that the separate actions for each is taken. For example, the preparation of the voter's list for election purposes, income tax assessment, recruitment of personnel, etc. are some of the areas where the census method is adopted. This method can be used where the population is comprised of heterogeneous items, i.e. different characteristics.

Though the census method provides a complete data of the population under study, it is very costly and time-consuming. Often, this method is dropped down because of these constraints and the sampling method, where certain items representative of the larger group, is selected to draw the conclusions.

Advantages and Disadvantages of Census Method

Following are the advantages and disadvantages of census method of data collection.

Advantages of Census Method

Intensive Study: Data collection through census method gives opportunity to the investigator to have an intensive study about a problem study about a problem. The investigator gathers a lot of knowledge through this method.

Intensive Study

Higher Degree of Accuracy

Higher Degree of Accuracy In this method there would be higher degree of accuracy in data. No other method is accurate like census method when the universe is small.

Higher Degree of Accuracy

Suitable for Heterogeneous Units

Suitable for Heterogeneous Units: This method is also applicable for units having heterogeneity or difference.

Indispensable in Some Cases

Indispensable in Some Cases: In certain cases this method is very important and suitable to be used for data collection. Without this method the study of a universe remains uncompleted.

Disadvantages of Census Method of Data Collection

Following are the demerits of census method.

Inconvenient

Inconvenient: This method is inconvenient because it take much time, money and set of industrial organizations.

Possible in Few Cases

Possible in Few Cases: This method is possible only in few cases of data collection.

Possible in Limited Circumstances

Possible in Limited Circumstances: This method is possible only in limited universe it requires not vast area to study.

Time and Labour Consuming

Time & Labour Consuming: It is much time and labour consuming. It requires a lot of people to involve in data collection.

More Statistical Errors

More Statistical Errors: Data collected through this method would have more statistical errors. Census method of data collection is helpful only in an area where inquiry is limited, more time and finance available and data have high degree of accuracy otherwise not.

Survey Method

A Survey is defined is a research method used for collecting data from a pre-defined group of respondents to gain information and insights on various topics of interest. Surveys have a variety of purposes and can be carried out in many ways depending on the methodology chosen and the objectives to be achieved.

The data is usually obtained through the use of standardized procedures whose purpose is to ensure that each respondent is able to answer the questions at a level playing field to avoid biased opinions that could influence the outcome of the research or study. A survey involves asking people for information through a questionnaire, which can be distributed on paper, although with the arrival of new technologies it is more common to distribute them using digital media such as social networks, email, QR codes or URLs.

Characteristics of a Survey

The need to observe or research facts about a situation leads us to conduct a survey. As we mentioned at the beginning, a survey is a method of gathering information.

Need to Conducting a Survey

First, a sample, also referred to as audience, is needed which should consist of a series of survey respondent's data with required demographic characteristics, who can relevantly answer your survey questions and provide the best insights. Better the quality of your survey sample, better will be your response quality and better your insights.

Surveys come in many different forms and have a variety of purposes, but they have common underlying characteristics. The basic characteristics of a survey are:

Survey Questions: Questionnaire for your Survey

Effective survey questions are the cornerstone for the success of any survey and subsequently, any research study. Whether it an email survey, SMS survey, web intercept survey or a mobile app survey, the single common denominator that determines how effectively you are able to collect accurate and complete survey responses is your survey questions and their types.

Multiple choice questions are the most common type of survey questions, in which, some of the popular question types are: dichotomous question, semantic differential scale question, rank order questions, rating scale questions and open ended questions.

To conduct a survey, it is important to plan the type of survey to ensure you get the optimum number of responses required for your survey. It could be a mix of interviews and survey questions or a questionnaire. Interviews could be telephone interviews, face-to-face interviews, online interviews and questionnaires can be mall surveys or web surveys. The underlying difference between a survey and a questionnaire is that a questionnaire may or may not be delivered in the form of a survey, but a survey always consists of a questionnaire.

Sample and Sample Determination

A sample is a selection of respondents from a population in such a manner that the sample represents the total population as closely as possible. Once you have determined your sample, the total number of individuals in that particular sample is the sample size.

Selecting a sample size depends on the end objective of your research study. It should consist of a series of survey respondents data with required demographic characteristics, who can relevantly answer your survey questions and provide the best insights. If the quality of the sample for survey is good, the response quality and insights are better.

Survey Methods

Survey methodology studies the in-depth sampling of individual units from a population and administering data collection techniques on that sample. It includes instruments or processes that ask different question types to a predefined sample, to conduct data-collection and increase the survey response rate.

The two distinctive member types are in a survey methodology are, professionals in the field that focus on empirical survey errors and others that work to design surveys and reduce them. it is therefore both a scientific field and a profession. The primary tasks of a survey methodologist while administering a survey is to identify and create samples, validate test questions, select the mode to administer questions and validate methods for data collection, statistical analysis and data reporting.

Survey Methods Based on Design

Surveys can be administered by the time they take to complete, the two types are:

- Cross-sectional studies: Cross-sectional study is defined as an observational research type that analyzes data of variables collected at one given point of time across a sample population. population or a pre-defined subset. This study type is also known as cross-sectional analysis, transverse study or prevalence study. The data gathered in a cross-sectional study is from people who are similar in all variables except the one variable which is under study. This variable remains constant throughout the cross-sectional study. This is unlike longitudinal study, where variables in the study can change over the course of research.

- Longitudinal studies: Longitudinal study is an observational study that employs continuous or repeated measures to follow particular individuals over prolonged period of time often years or decades. Longitudinal study collects data that is either qualitative or quantitative in nature. In longitudinal study a survey creator is not interfering with survey respondents. Survey respondents are observed over a period of time ranging from months to even decades to observe any changes in them or their attitude. For example, a researcher wants to find out which disease affects young boys (in the age group of 10-15) then the researcher will observe the individuals over that period to collect meaningful data.

Survey Methods based on Distribution

There are different ways of survey distribution. Some of the most commonly used methods are:

- Email: Sending out an email is the easiest way of conducting a survey. The respondents are targeted and there is higher chance of response due to the the respondents already knowing about your brand. You can use the QuestionPro email management feature to send out and collect survey responses.
- Buy respondents: Buying a sample helps achieve a lot of the response criteria because the people who are being asked to respond have signed up to do so and the qualifying criteria for the research study is met.
- Embed survey in website: Embedding a survey in a website ensures that the number of responses is very high. This can be done while the person enters the website or is exiting it. A non intrusive method of collecting feedback is important to achieve a higher number of responses. The responses received are also honest due to the high brand recall value and the responses are quick to collect and analyze due to them being in a digital format.
- Post to social network: Posting on social networks is another effective way of collecting responses. The survey can be posted as a link and people that follow the brand can take a survey. This method is used when there is no upper cap on the number of survey responses required and is the easiest and fastest way of eliciting responses.
- QR code: QuestionPro QR codes store the URL for the survey. You can print/publish this code in magazines, on signs, business cards, or on just about any object/medium. Users with a camera phone equipped with the correct reader application can scan the image of the QR Code to open the survey in the phone's browser.
- QuestionPro App: The QuestionPro App allows to quickly circulate surveys and the responses can be collected both online and offline.
- API: You can use the API integration of the QuestionPro platform for potential respondents to take your survey.

- SMS: Using SMS surveys are another quick way to collect feedback. This method can be used in the case of quick responses and when the survey is simple, straightforward and not too long. This method is used to increase the open and response rate of collecting feedback.

Surveys can be distributed using one, some or a mix of the above methods depending on the basis of the research objective and the resources being used for any particular survey. Many factors play a part in the mode of distribution of surveys like cost, research study type, flexibility of questions, time to collect responses, statistical analysis to be run on data and willingness of the respondent to take part in the survey.

You can conduct a telephone or email survey and then make a selection of respondents for a face-to-face interview. Survey data are sometimes also obtained through questionnaires filled out by respondents in groups, for example, a school class or a group of shoppers in a shopping center.

You can also classify the surveys by their content, being able to use open or closed questions to know, for example, opinions, attitudes, details of a fact, habits, experiences for a later classification and analysis of the obtained results.

Online Survey

An online survey is set of structured questions that the respondent completes over the internet, generally through filling out a form. An online survey is an easier way to reach out to the respondents as it is less time consuming than the traditional way of gathering information through one to one interaction and also less expensive.

Online surveys can differ in length and format used. The data is collected and stored in a database which is later evaluated by an expert in the field.

As an incentive for respondents to fill out online surveys, businesses offer them with rewards like gift cards, reward points that they can redeem for goods or services later, free airline miles, discount at gas stations etc.

Online surveys with rewards is a win-win situation for both, businesses and respondents. The businesses or organizations get valuable data from a controlled environment, for the market research and for the respondents it's a humble gesture from businesses to participate in the survey and spend their valuable time.

Advantages of Online Survey

Online surveys are one of the less time consuming methods to reach out to your target audiences, but that's not the only advantage of online survey, let us understand what are the other advantages of using online survey:

- Accuracy: In online surveys, usually the margin of error is reduced, as the respondents register their responses by easy selection buttons. Tradition methods require human interference and according to a study human interference increase the margin of error by 10%.

- Easy and quick to analyze: Since all the responses are registered online, it is extremely easy and quick to analyze the data in real time. It is also quick to draw inferences and share the result.
- Ease of participation: In this new age technology oriented universe, most people on this planet have access to the internet. Respondents prefer receiving the survey over the email. Ease of participation greatly increases as the respondents can choose a suitable time and place, according to their convenience to register their responses.
- Great branding exercise: While designing the survey online, organizations or businesses have this opportunity to design their survey to align to their brand. Using logos and similar brand language (color and fonts) in the survey, gives an advantage to the businesses as respondents are able to connect better with the brand.
- Respondents can be honest and flexible at the same time: According to a study researchers have found an increased participation by respondents when deployed with online surveys rather than answering lengthy questions. By designing surveys that ask relevant questions, respondents are honest with their answers and can skip the questions or respondent to a more neutral option in the survey, increasing their flexibility to respond.

Survey Examples

It is important for a researcher to be able to conduct surveys using the right questions and the right medium to administer and track responses. QuestionPro is a platform that helps not only to create but also to deploy different types of survey. We have more than 300+ types of survey template and survey examples including:

- Customer Satisfaction (CSAT) + Net Promoter Score (NPS) Survey: We hear this time and again that customer is king; and it is true. A satisfied customer is a customer that helps your brand and organization grow, through direct means as well as being an advocate for your brand. This survey template talks about the goodwill your brand has created and how referenceable it is.
- Company Communications Evaluation Survey Template: This survey example is usually used to analyze employee perspective about what is important to be communicated regularly, topics that should be included in the newsletter, updates on the bulletin board, the efficiency of an organization's management in communication etc.
- Hardware Product Evaluation Survey Template: Improving hardware product features isn't a very easy proposition due to a lot of elements like raw materials, supply chain and manufacturing lines getting affected by it. Hence, while eliciting feedback for hardware, it is important to be as objective as possible. This survey helps understand exactly what product innovation is required.
- Strategic Planning Survey: Innovation is important to any organization's product or service lines. Hence, implementing customer support and making product or service tweaks when required are extremely important for the sustenance and growth of an organization. This survey template helps organizations chalk out their business strategy.

- Business Demographic Survey: This survey template is designed to carry demographic questions and examples that help gain information on occupation, primary area of business, job function and description, organization's gross income etc.
- Course Evaluation Survey: This survey template helps educational institutions conduct period feedback on their course and if students find it helpful or not, if it's stimulating enough and students see this is as value for money along with accentuated learning.

References

- Primary-and-secondary-data-in-statistics: itfeature.com, Retrieved 23, May 2020
- What-is-census-method-of-data-collection-advantages-disadvantages, social-research-methodology: studylecturenotes.com, Retrieved 27, August 2020
- Statistical-data-analysis: statisticssolutions.com, Retrieved 08, January 2020
- Types-of-data-primary-and-secondary-data: communitymedicine4asses.com, Retrieved 21, July 2020
- Difference-between-primary-and-secondary-data: keydifferences.com, Retrieved 18, March 2020

6

Types of Regression Techniques

The set of statistical methods used to estimate the relationship between dependent and independent variables is known as regression analysis. This chapter discusses in detail topics such as linear regression analysis, nonlinear regression, nonparametric regression, semiparametric regression and regression models.

The Regression Analysis is a statistical tool used to determine the probable change in one variable for the given amount of change in another. This means, the value of the unknown variable can be estimated from the known value of another variable.

The degree to which the variables are correlated to each other depends on the Regression Line. The regression line is a single line that best fits the data, i.e. all the points plotted are connected via a line in the manner that the distance from the line to the points is the smallest.

The regression also tells about the relationship between the two or more variables, then what is the difference between regression and correlation? Well, there are two important points of differences between Correlation and Regression. These are:

- The Correlation Coefficient measures the “degree of relationship ”between variables, say X and Y whereas the Regression Analysis studies the “nature of relationship” between the variables.
- Correlation coefficicnt does not clearly indicate the cause-and-effect relationship between the variables, i.e. it cannot be said with certainty that one variable is the cause, and the other is the effect. Whereas, the Regression Analysis clearly indicates the cause-and-effect relationship between the variables.

The regression analysis is widely used in all the scientific disciplines. In economics, it plays a significant role in measuring or estimating the relationship among the economic variables. For example, the two variables – price (X) and demand (Y) are closely related to each other, so we can find out the probable value of X from the given value of Y and similarly the probable value of Y can be found out from the given value of X.

Types of Regression Analysis

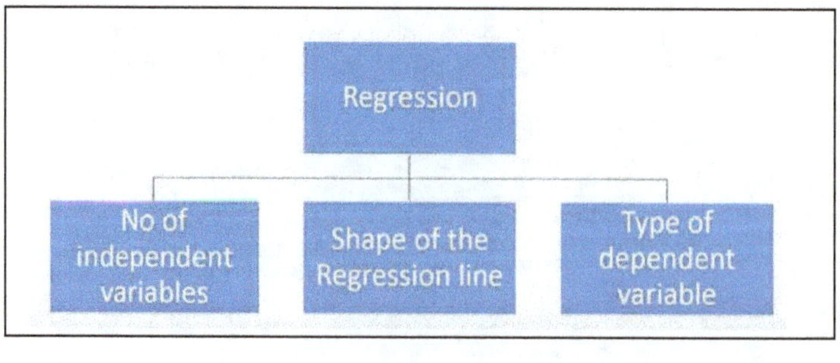

There are various kinds of regression techniques available to make predictions. These techniques are mostly driven by three metrics (number of independent variables, type of dependent variables and shape of regression line).

Linear Regression Analysis

Linear regression is a basic and commonly used type of predictive analysis. The overall idea of regression is to examine two things:

(1) Does a set of predictor variables do a good job in predicting an outcome (dependent) variable?

(2) Which variables in particular are significant predictors of the outcome variable, and in what way do they–indicated by the magnitude and sign of the beta estimates–impact the outcome variable? These regression estimates are used to explain the relationship between one dependent variable and one or more independent variables. The simplest form of the regression equation with one dependent and one independent variable is defined by the formula y = c + b*x, where y = estimated dependent variable score, c = constant, b = regression coefficient, and x = score on the independent variable.

Naming the Variables:

There are many names for a regression's dependent variable. It may be called an outcome variable, criterion variable, endogenous variable, or regressand. The independent variables can be called exogenous variables, predictor variables, or regressors.

Three major uses for regression analysis are (1) determining the strength of predictors, (2) forecasting an effect, and (3) trend forecasting.

First, the regression might be used to identify the strength of the effect that the independent variable(s) have on a dependent variable. Typical questions are what is the strength of relationship between dose and effect, sales and marketing spending, or age and income.

Second, it can be used to forecast effects or impact of changes. That is, the regression analysis helps us to understand how much the dependent variable changes with a change in one or more independent variables.

Third, regression analysis predicts trends and future values. The regression analysis can be used to get point estimates. A typical question is, "what will the price of gold be in 6 months?"

Types of Linear Regression

Simple Linear Regression

Simple linear regression is a statistical method that allows us to summarize and study relationships between two continuous (quantitative) variables:

- One variable, denoted x, is regarded as the predictor, explanatory, or independent variable.
- The other variable, denoted y, is regarded as the response, outcome, or dependent variable.

Because the other terms are used less frequently today, we'll use the "predictor" and "response" terms to refer to the variables encountered in this course. The other terms are mentioned only to make you aware of them should you encounter them in other arenas. Simple linear regression gets its adjective "simple," because it concerns the study of only one predictor variable.

Types of Relationships

Before proceeding, we must clarify what types of relationships we won't study in this course, namely, deterministic (or functional) relationships. Here is an example of a deterministic relationship.

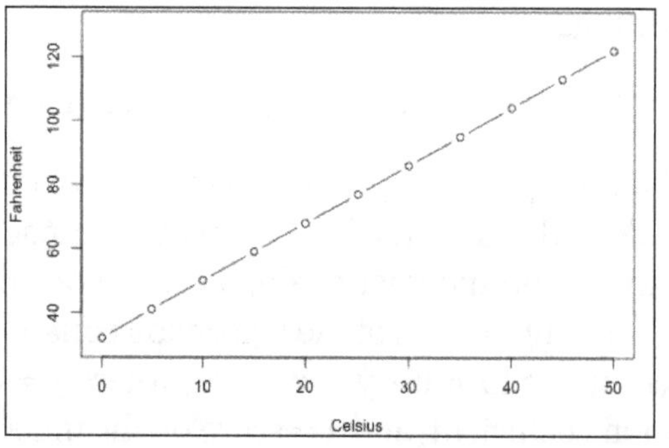

Note that the observed (x, y) data points fall directly on a line. As you may remember, the relationship between degrees Fahrenheit and degrees Celsius is known to be:

$$\text{Fahr}=\frac{9}{5}\text{Cels}+32$$

That is, if you know the temperature in degrees Celsius, you can use this equation to determine the temperature in degrees Fahrenheit *exactly*.

Here are some examples of other deterministic relationships that students from previous semesters have shared:

- Circumference = $\pi \times$ diameter.
- Hooke's Law: $Y = \alpha + \beta X$, where Y = amount of stretch in a spring, and X = applied weight.
- Ohm's Law: $I = V/r$, where V = voltage applied, r = resistance, and I = current.
- Boyle's Law: For a constant temperature, $P = \alpha/V$, where P = pressure, α = constant for each gas, and V = volume of gas.

For each of these deterministic relationships, the equation *exactly* describes the relationship between the two variables. This course does not examine deterministic relationships. Instead, we are interested in statistical relationships, in which the relationship between the variables is not perfect.

Multiple Linear Regression Model

A linear regression model that contains more than one predictor variable is called a *multiple linear regression model.* The following model is a multiple linear regression model with two predictor variables, x_1 and x_2.

$$Y = \beta_0 + \beta_1 x_1 + \beta_2 x_2 + \in$$

The model is linear because it is linear in the parameters β_0, β_1 and β_2. The model describes a plane in the three-dimensional space of Y, x_1 and x_2. The parameter β_0 is the intercept of this plane. Parameters β_1 and β_2 are referred to as *partial regression coefficients*. Parameter β_1 represents the change in the mean response corresponding to a unit change in x_1 when x_2 is held constant. Parameter β_2 represents the change in the mean response corresponding to a unit change in x_2 when x_1 is held constant. Consider the following example of a multiple linear regression model with two predictor variables, x_1 and x_2:

$$Y = 30 + 5x_1 + 7x_2 + \in$$

This regression model is a first order multiple linear regression model. This is because the maximum power of the variables in the model is 1. (The regression plane corresponding to this model is shown in the figure below.) Also shown is an observed data point and the corresponding random error, $\in$. The true regression model is usually never known (and therefore the values of the random error terms corresponding to observed data points remain unknown). However, the regression model can be estimated by calculating the parameters of the model for an observed data set.

One of the following figures shows the contour plot for the regression model the above equation. The contour plot shows lines of constant mean response values as a function of x_1 and x_2. The contour lines for the given regression model are straight lines as seen on the plot. Straight contour lines result for first order regression models with no interaction terms.

A linear regression model may also take the following form:

$$Y = \beta_0 + \beta_1 x_1 + \beta_2 x_2 + \beta_{12} x_1 x_2 + \in$$

A cross-product term, $x_1 x_2$, is included in the model. This term represents an interaction effect between the two variables x_1 and x_2. Interaction means that the effect produced by a change in the predictor variable on the response depends on the level of the other predictor variable(s). As an example of a linear regression model with interaction, consider the model given by the equation $Y = 30 + 5x_1 + 7x_2 + 3x_1 x_2 + \in$. The regression plane and contour plot for this model are shown in the following two figures, respectively.

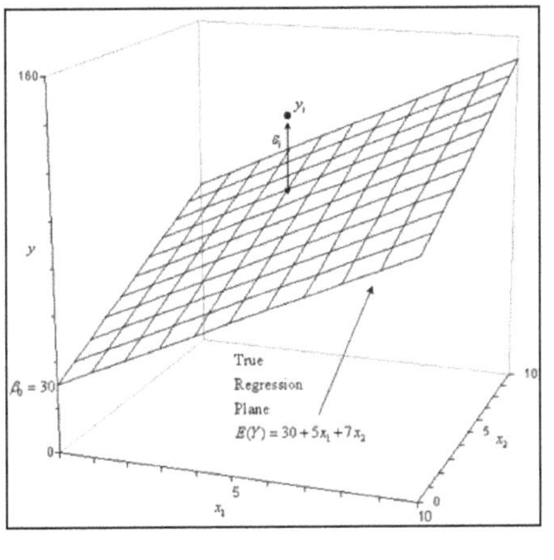

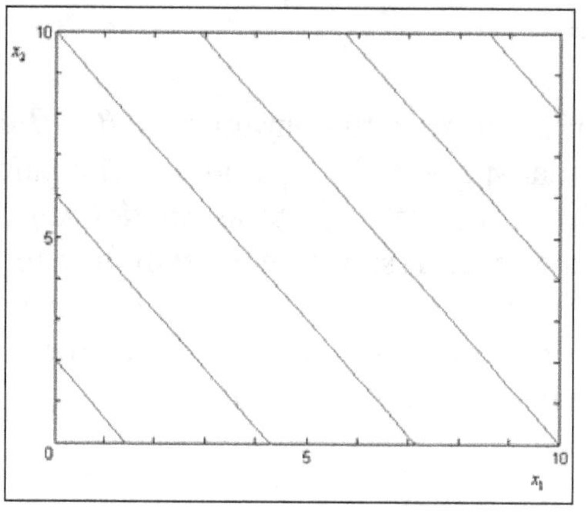

Now consider the regression model shown next:

$$Y = \beta_0 + \beta_1 x_1 + \beta_2 x_1^2 + \beta_3 x_1^3 + \in$$

This model is also a linear regression model and is referred to as a *polynomial regression model.* Polynomial regression models contain squared and higher order terms of the predictor variables making the response surface curvilinear. As an example of a polynomial regression model with an interaction term consider the following equation:

$$Y = 500 + 5x_1 + 7x_2 - 3x_1^2 - 5x_2^2 + 3x_1 x_2 + \in$$

This model is a *second order* model because the maximum power of the terms in the model is two. The regression surface for this model is shown in the following figure. Such regression models are used in RSM to find the optimum value of the response Y. Notice that, although the shape of the regression surface is curvilinear, the regression model is still linear because the model is linear in the parameters. The contour plot for this model is shown in the second of the following two figures:

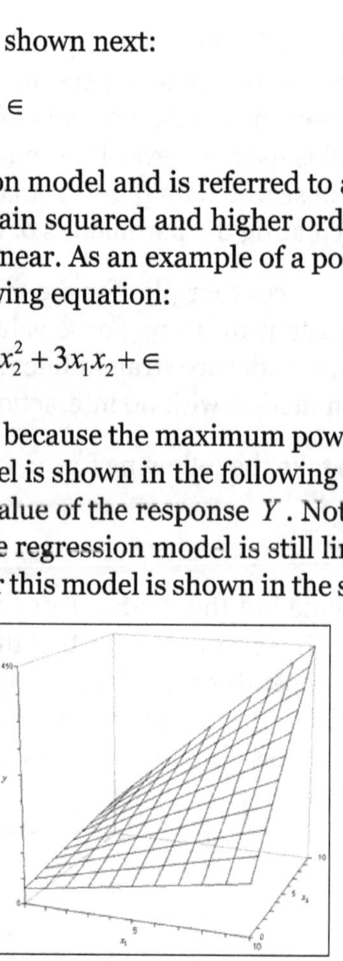

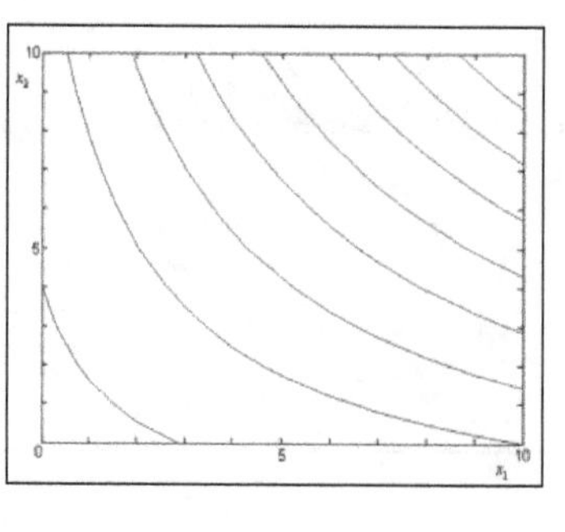

All multiple linear regression models can be expressed in the following general form:

$$Y = \beta_0 + \beta_1 x_1 + \beta_2 x_2 + \ldots + \beta_k x_k + \in$$

where k denotes the number of terms in the model. For example, the model can be written in the general form using $x_3 = x_1^2$, $x_4 = x_2^3$ and $x_5 = x_1 x_2$ as follows:

$$Y = 500 + 5x_1 + 7x_2 - 3x_3 - 5x_4 + 3x_5 + \in$$

Logistic Regression

In statistics, the logistic model (or logit model) is a widely used statistical model that, in its basic form, uses a logistic function to model a binary dependent variable; many more complex extensions exist. In regression analysis, logistic regression (or logit regression) is estimating the parameters of a logistic model; it is a form of binomial regression. Mathematically, a binary logistic model has a dependent variable with two possible values, such as pass/fail, win/lose, alive/dead or healthy/sick; these are represented by an indicator variable, where the two values are labeled "0" and "1". In the logistic model, the log-odds (the logarithm of the odds) for the value labeled "1" is a linear combination of one or more independent variables ("predictors"); the independent variables can each be a binary variable (two classes, coded by an indicator variable) or a continuous variable (any real value). The corresponding probability of the value labeled "1" can vary between 0 (certainly the value "0") and 1 (certainly the value "1"), hence the labeling; the function that converts log-odds to probability is the logistic function, hence the name. The unit of measurement for the log-odds scale is called a *logit*, from *logistic unit*, hence the alternative names. Analogous models with a different sigmoid function instead of the logistic function can also be used, such as the probit model; the defining characteristic of the logistic model is that increasing one of the independent variables multiplicatively scales the odds of the given outcome at a *constant* rate, with each dependent variable having its own parameter; for a binary independent variable this generalizes the odds ratio.

Logistic regression was developed by statistician David Cox in 1958. The binary logistic regression model has extensions to more than two levels of the dependent variable: categorical outputs with more than two values are modelled by multinomial logistic regression, and if the multiple categories are ordered, by ordinal logistic regression, for example the proportional odds ordinal logistic model. The model itself simply models probability of output in terms of input, and does not perform statistical classification (it is not a classifier), though it can be used to make a classifier, for instance by choosing a cutoff value and classifying inputs with probability greater than the cutoff as one class, below the cutoff as the other; this is a common way to make a binary classifier. The coefficients are generally not computed by a closed-form expression, unlike linear least squares.

Applications

Logistic regression is used in various fields, including machine learning, most medical fields, and social sciences. For example, the Trauma and Injury Severity Score (TRISS), which is widely used to predict mortality in injured patients, was originally developed by Boyd et al. using logistic regression. Many other medical scales used to assess severity of a patient have been developed using logistic regression. Logistic regression may be used to predict the risk of developing a given disease (e.g. diabetes; coronary heart disease), based on observed characteristics of the patient (age, sex, body mass index, results of various blood tests, etc.). Another example might be to predict whether an Indian voter will vote BJP or Trinamool Congress or Left Front or Congress, based on age, income, sex, race, state of residence, votes in previous elections, etc. The technique can also be used in engineering, especially for predicting the probability of failure of a given process, system or product. It is also used in marketing applications such as prediction of a customer's propensity to purchase a product or halt a subscription, etc. In economics it can be used to predict the likelihood of a person's choosing to be in the labor force, and a business application would be to predict the likelihood of a homeowner defaulting on a mortgage. Conditional random fields, an extension of logistic regression to sequential data, are used in natural language processing.

Examples

Logistic Model

One may begin to understand logistic regression by first considering a logistic model with given parameters, then seeing how the coefficients can be estimated ("regressed") from data. Consider a model with two predictors, x_1 and x_2; these may be continuous variables (taking a real number as value), or indicator functions for binary variables (taking value 0 or 1). Then the general form of the log-odds (here denoted by *l*) is:

$$l = \beta_0 + \beta_1 x_1 + \beta_2 x_2,$$

where the coefficients β_i are the parameters of the model. Note that this is a linear model: the log-odds *l* are a linear combination of the predictors x_1 and x_2, including a constant term β_0. The corresponding odds are the exponent:

$$o = b^{\beta_0 + \beta_1 x_1 + \beta_2 x_2}$$

where *b* is the base of the logarithm and exponent. This is now a non-linear model, since the odds are not a linear combination of the predictors.

Odds of $o = o:1$ for an event are the same as probability of $o/(o+1)$, since happening *o* of the time and not happening 1 time corresponds to happening *o* times out of a total of *o* + 1 events, so the corresponding probability is:

$$p = \frac{b^{\beta_0 + \beta_1 x_1 + \beta_2 x_2}}{b^{\beta_0 + \beta_1 x_1 + \beta_2 x_2} + 1} = \frac{1}{1 + b^{-(\beta_0 + \beta_1 x_1 + \beta_2 x_2)}}$$

The base *b* is usually taken to be *e*, but for exposition here use 10, so the odds are more understandable. Suppose the coefficients are $-3, 1, 2$, so the model is:

$$l = -3 + 1 \cdot x_1 + 2 \cdot x_2,$$

This can be interpreted as follows:

- $\beta_0 = -3$ is the *y*-intercept, and can thus be interpreted as the log-odds when the predictors are all zero; here the log-odds are -3, so the odds for are $10^{-3} = 10^{-3} : 1 = 1 : 1000$, and the probability is $1/(1000+1) = 1/1001 \approx 0.001 = 0.1\%$.

- Alternatively, grouping the weighted predictors as a single contribution to the total logit, $l' = 1 \cdot x_1 + 2 \cdot x_2$, so $l = l' - 3$, the quantity $-\beta_0$ is the x-intercept, and can be interpreted as the weighted predictors corresponding to even odds: if $l' = 1 \cdot x_1 + 2 \cdot x_2 = 3$, then the log-odds are 0, the odds for are $10^0 = 10^0 : 1 = 1 : 1$, and the probability is $1/(1+1) = 1/2 = 50\%$.
- $\beta_1 = 1$ means that increasing x_1 by 1 increases the log-odds by $1 \cdot 1$, so it *multiplies* the odds by 10 10; this is sometimes referred to as the "effect" of the predictor x_1.
- $\beta_2 = 2$ means that increasing x_2 by 1 increases the log-odds by $2 \cdot 1$, so it *multiplies* the odds by $10^2 = 100$ Thus the effect of x_2 on the log-odds is twice as great as the effect of x_1.

Such a model can be used for various purposes. For example, given an individual datum with values of predictors x_1, x_2, one can estimate the log-odds (hence odds and probability) of the outcome by putting in the values in the formula. Alternatively, consider comparing two medical treatments, one of which decreases x_1 by 3 and the other that decreases x_2 by 2. Assuming this model is valid, the first treatment reduces the log-odds of the outcome by $1 \cdot 3 = 3$, so it reduces the odds by a factor of $10^3 = 1000$ but the other treatment reduces the log-odds by $2 \cdot 2 = 4$, so it reduces the odds by a factor of $10^4 = 10000$, and thus is more effective, all else equal.

In order to estimate the parameters of such a logistic model and compute how well it fits the data, one must do a logistic regression.

Probability of Passing an Exam Versus Hours of Study

Suppose we wish to answer the following question:

A group of 20 students spend between 0 and 6 hours studying for an exam. How does the number of hours spent studying affect the probability that the student will pass the exam?

The reason for using logistic regression for this problem is that the values of the dependent variable, pass and fail, while represented by "1" and "0", are not cardinal numbers. If the problem was changed so that pass/fail was replaced with the grade 0–100 (cardinal numbers), then simple regression analysis could be used.

The table shows the number of hours each student spent studying, and whether they passed (1) or failed (0).

Hours	0.50	0.75	1.00	1.25	1.50	1.75	1.75	2.00	2.25	2.50	2.75	3.00	3.25	3.50	4.00	4.25	4.50	4.75	5.00	5.50
Pass	0	0	0	0	0	0	1	0	1	0	1	0	1	0	1	1	1	1	1	1

The graph shows the probability of passing the exam versus the number of hours studying, with the logistic regression curve fitted to the data.

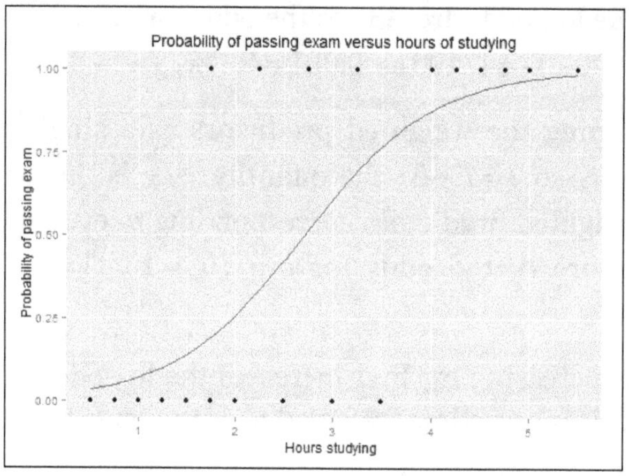

Graph of a logistic regression curve showing probability of passing an exam versus hours studying

The logistic regression analysis gives the following output.

	Coefficient	Std.Error	z-value	P-value (Wald)
Intercept	−4.0777	1.7610	−2.316	0.0206
Hours	1.5046	0.6287	2.393	0.0167

The output indicates that hours studying is significantly associated with the probability of passing the exam ($p = 0.0167$, Wald test). The output also provides the coefficients for Intercept $= -4.0777$ and Hours $= 1.5046$. These coefficients are entered in the logistic regression equation to estimate the odds (equivalently, probability) of passing the exam:

$$\text{Log-odds of passing exam} = 1.5046 \cdot \text{Hours} - 4.0777 = 1.5046 \cdot (\text{Hours} - 2.71)$$

$$\text{Odds of passing exam} = \exp(1.5046 \cdot \text{Hours} - 4.0777) = \exp(1.5046 \cdot (\text{Hours} - 2.71))$$

$$\text{Probability of passing exam} = \frac{1}{1 + \exp(-(1.5046 \cdot \text{Hours} - 4.0777))}$$

One additional hour of study is estimated to increase log-odds of passing by 1.5046, so multiplying odds of passing by $\exp(1.5046) \approx 4.5$. The form with the x-intercept (2.71) shows that this estimates even odds (log-odds 0, odds 1, probability 1/2) for a student who studies 2.71 hours.

For example, for a student who studies 2 hours, entering the value Hours $= 2$ in the equation gives the estimated probability of passing the exam of 0.26:

$$\text{Probability of passing exam} = \frac{1}{1 + \exp(-(1.5046 \cdot 2 - 4.0777))} = 0.26$$

Similarly, for a student who studies 4 hours, the estimated probability of passing the exam is 0.87:

$$\text{Probability of passing exam} = \frac{1}{1 + \exp(-(1.5046 \cdot 4 - 4.0777))} = 0.87$$

This table shows the probability of passing the exam for several values of hours studying.

Hours of study	Passing exam		
	Log-odds	Odds	Probability
1	−2.57	0.076 ≈ 1:13.1	0.07
2	−1.07	0.34 ≈ 1:2.91	0.26
3	0.44	1.55	0.61
4	1.94	6.96	0.87
5	3.45	31.4	0.97

The output from the logistic regression analysis gives a p-value of $p = 0.0167$, which is based on the Wald z-score. Rather than the Wald method, the recommended method to calculate the p-value for logistic regression is the likelihood-ratio test (LRT), which for this data gives $p = 0.0006$.

Ordinal regression is a statistical technique that is used to predict behavior of ordinal level dependent variables with a set of independent variables. The dependent variable is the order response category variable and the independent variable may be categorical or continuous.

Ordered Regression

Assumptions:

- One dependent variable, we cannot use multiple dependent variables.
- Parallel lines assumption: There is one regression equation for each category except the last category. The last category probability can be predicted as 1-second last category probability.
- Adequate cell count: As per the rule of thumb, 80% of cells must have more than 5 counts. No cell should have Zero count. The greater the cell with less count, the less reliable the chi-square test will be.

Multinomial Logistic Regression

Multinomial logistic regression is used when you have a categorical dependent variable with two or more unordered levels (i.e. two or more discrete outcomes). It is practically identical to logistic regression, except that you have multiple possible outcomes instead of just one.

For example, children's food choices are influenced by their parents' choices and the children's pastimes (e.g. sports enthusiast vs. gamer). You could study the relationship between a child's food choices with their parents' choices and children's pastimes. The dependent variable levels would be the different food choices (fast food, healthy choices, protein packed, vegan etc.). Or you might study how workers' education levels and time on the job affect promotions. The independent variables would be education levels and time on the job, and the levels of the dependent variable might be promotion to team-leader roles, sales positions, or management positions.

One level of the dependent variable is chosen as the *reference category*. This is typically the most common or the most frequent category. In the first example above, this might be "fast food". The probability of being in any of the other categories is compared to the probability of being in the reference category. These relative probabilities are the predicted log odds.

Running Multinomial Logistic Regression

This type of regression is usually performed with software. Essentially, the software will run a series of individual binomial logistic regressions for M – 1 categories (one calculation for each category, minus the reference category). When M = 2, multinomial logistic regression, ordered logistic regression, and logistic regression are equal.

Before the advent of computer software, you would have run these individual regressions and then compared the results. The software takes away that chore, and estimates parameters simultaneously, resulting in more efficiency.

Assumptions include:

- The model is specified correctly with no extraneous variables.
- Cases are independent.
- There is no multicollinearity between the independent variables.

Interpreting Results

Multinomial logistic regression works the same way as other types of regression: you're looking for a relationship between the independent and dependent variables. The output will give you sets of coefficients for each variable. The output for each software package will vary.

Nonlinear Regression

Nonlinear regression uses nonlinear regression equations, which take the form:

$$Y = f(X, \beta) + \varepsilon$$

Where:

- X = a vector of p predictors,
- β = a vector of k parameters,
- f(-) = a known regression function,
- ε = an error term.

The formal definition is that if your regression equation looks like the one above, it's nonlinear regression. However, this is actually a lot more difficult than it sounds. Take the following nonlinear regression equations:

- The Michaelis -Menten model: $f(x,\beta)=(\beta_1 x)/(\beta_2+x)$.
- $Y = \beta_0 + (0.4 - \beta_0)e^{-\beta_1}(x_i-5) + \varepsilon_i$.

These both meet the requirement of fitting the form $Y = f(X,\beta) + \varepsilon$, but that isn't immediately obvious without some in-depth knowledge of algebra and regression analysis.

The good news is there is a much simpler, more intuitive definition of nonlinear regression:

If your model uses an equation in the form $Y = a_0 + b_1X_1$, it's a linear regression model. If not, it's nonlinear.

It's *much* easier to spot a linear regression equation, as it's *always* going to take the form $Y = a_0 + b_1X_1$ *.

Nonparametric Regression

Nonparametric regression, like linear regression, estimates mean outcomes for a given set of covariates. Unlike linear regression, nonparametric regression is agnostic about the functional form between the outcome and the covariates and is therefore not subject to misspecification error.

In nonparametric regression, you do not specify the functional form. You specify the dependent variable—the outcome—and the covariates. You specify y, x_1, x_2, and x_3 to fit.

$$y = g(x_1, x_2, x_3) + \epsilon$$

The method does not assume that g()is linear; it could just as well be

$$y = \beta_1 x_1 + \beta_2 x_2^2 + \beta_3 x_1^3 x_2 + \beta_4 x_3 + \epsilon$$

The method does not even assume the function is linear in the parameters. It could just as well be

$$y = \beta_1 x_1^{\beta 2} + \cos(x_2 x_3) + \epsilon$$

Kernel Regression

Kernel regression is a modeling tool which belongs to the family of smoothing methods. Unlike linear regression which is both used to explain phenomena and for prediction (understanding a phenomenon to be able to predict it afterwards), Kernel regression is mostly used for prediction. The structure of the model is variable and complex, the latter working like a filter or black box. There are many variations of Kernel regression in existence.

As with any modeling method, a learning sample of size nlearn is used to estimate the parameters of the model. A sample of size nvalid can then be used to evaluate the quality of the model. Lastly, the model can be applied to a prediction sample of size npred, for which the values of the dependent variable Y are unknown.

The characteristics of Kernel Regression are:

- The use of a kernel function, to weigh the observations of the learning sample, depending on their "distance" from the predicted observation.

  The kernel functions available in XLSTAT are:

  - Uniform
  - Triangle
  - Epanechnikov
  - Quartic
  - Triweight
  - Tricube
  - Gaussian
  - Cosine

- The bandwidth associated to each variable. It is involved in calculating the kernel and the weights of the observations, and differentiates or rescales the relative weights of the variables while at the same time reducing or augmenting the impact of observations of the learning sample, depending on how far they are from the observation to predict.
- The polynomial degree used when fitting the model to the observations of the learning sample. Two strategies are suggested in order to restrict the size of the learning sample taken into account for the estimation of the parameters of the polynomial: Moving window and k nearest neighbors.

Lowess Regression

Locally weighted regression and smoothing scatter plots or LOWESS regression was introduced to create smooth curves through scattergrams.

LOWESS regression is very similar to Kernel regression as it is also based on polynomial regression and requires a kernel function to weight the observations.

Gaussian Process Regression or Kriging

In Gaussian process regression, also known as Kriging, a Gaussian prior is assumed for the regression curve. The errors are assumed to have a multivariate normal distribution and the regression curve is estimated by its posterior mode. The Gaussian prior may depend on unknown hyperparameters, which are usually estimated via empirical Bayes.

Smoothing splines have an interpretation as the posterior mode of a Gaussian process regression.

Nonparametric Multiplicative Regression

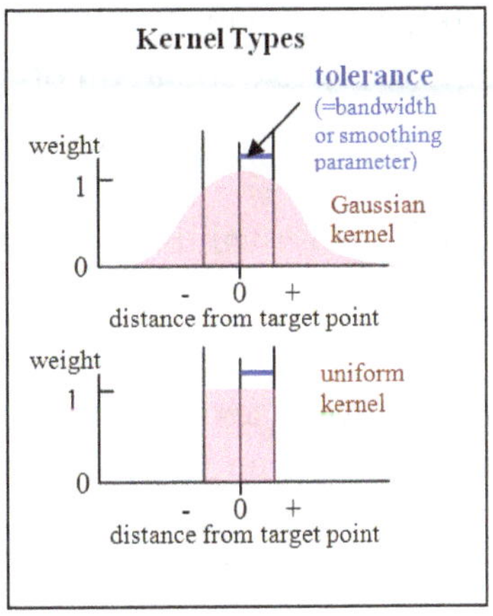

Two kinds of kernels used with kernel smoothers for nonparametric regression.

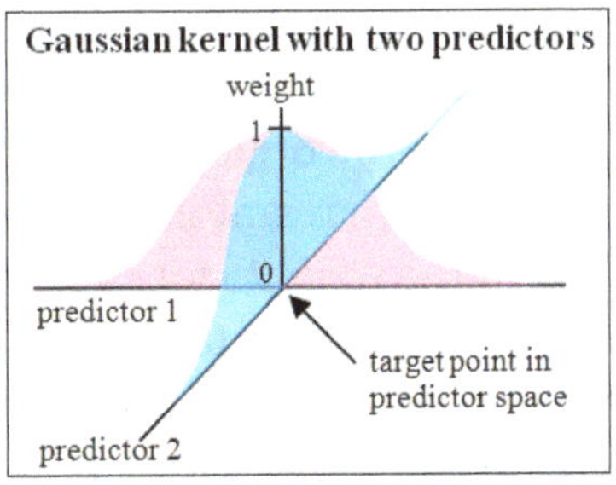

Use of Gaussian kernels for nonparametric multiplicative regression with two predictors. The weights from the kernel function for each predictor are multiplied to obtain a weight for a given data point in estimating a response variable (dependent variable) at a target point in the predictor space.

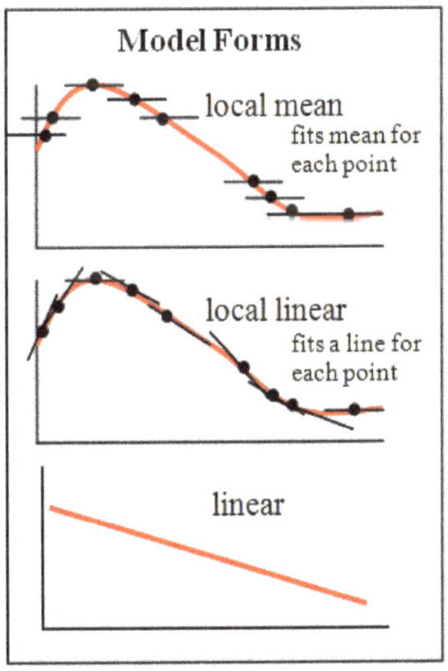

Two commonly used forms of a local model used in nonparametric regression, contrasted with a simple linear model.

Nonparametric multiplicative regression (NPMR) is a form of nonparametric regression based on multiplicative kernel estimation. Like other regression methods, the goal is to estimate a response (dependent variable) based on one or more predictors (independent variables). NPMR can be a good choice for a regression method if the following are true:

- The shape of the response surface is unknown.
- The predictors are likely to interact in producing the response; in other words, the shape of the response to one predictor is likely to depend on other predictors.
- The response is either a quantitative or binary (0/1) variable.

This is a smoothing technique that can be cross-validated and applied in a predictive way.

NPMR Behaves Like an Organism

NPMR has been useful for modeling the response of an organism to its environment. Organismal response to environment tends to be nonlinear and have complex interactions among predictors. NPMR allows you to model automatically the complex interactions among predictors in much the same way that organisms integrate the numerous factors affecting their performance.

A key biological feature of an NPMR model is that failure of an organism to tolerate any single dimension of the predictor space results in overall failure of the organism. For example, assume that a plant needs a certain range of moisture in a particular temperature range. If either temperature or moisture fall outside the tolerance of the organism, then the organism dies. If it is too hot, then no amount of moisture can compensate to result in survival of the plant. Mathematically this works with NPMR because the product of the weights for the target point is zero or near zero if any of the weights for individual predictors (moisture or temperature) are zero or near zero. Note further that in this simple example, the second condition listed above is probably true: the response of the plant to moisture probably depends on temperature and vice versa.

Optimizing the selection of predictors and their smoothing parameters in a multiplicative model is computationally intensive. With a large pool of predictors, the computer must search through a huge number of potential models in search for the best model. The best model has the best fit, subject to overfitting constraints or penalties.

Local Model

NPMR can be applied with several different kinds of local models. By "local model" we mean the way that data points near a target point in the predictor space are combined to produce an estimate for the target point. The most common choices for the local models are the local mean estimator, a local linear estimator, or a local logistic estimator. In each case the weights can be extended multiplicatively to multiple dimensions.

In words, the estimate of the response is a local estimate (for example a local mean) of the observed values, each value weighted by its proximity to the target point in the predictor space, the weights being the product of weights for individual predictors. The model allows interactions, because weights for individual predictors are combined by multiplication rather than addition.

Overfitting Controls

Understanding and using these controls on overfitting is essential to effective modeling with nonparametric regression. Nonparametric regression models can become overfit either by including too many predictors or by using small smoothing parameters (also known as bandwidth or tolerance). This can make a big difference with special problems, such as small data sets or clumped distributions along predictor variables.

The methods for controlling overfitting differ between NPMR and the generalized linear modeling (GLMs). The most popular overfitting controls for GLMs are the Akaike information criterion (AIC) and the Bayesian information criterion (BIC) for model selection. The AIC and BIC depend on the number of parameters in a model. Because NPMR models do not have explicit parameters as such, these are not directly applicable to NPMR models. Instead, one can control overfitting by setting a minimum average neighborhood size, minimum data:predictor ratio, and a minimum improvement required to add a predictor to a model.

Nonparametric regression models sometimes use an AIC based on the "effective number of parameters". This penalizes a measure of fit by the trace of the smoothing matrix—essentially how much each data point contributes to estimating itself, summed across all data points. If, however, you use leave-one-out cross validation in the model fitting phase, the trace of the smoothing matrix is always zero, corresponding to zero parameters for the AIC. Thus, NPMR with cross-validation in the model fitting phase already penalizes the measure of fit, such that the error rate of the training data set is expected to approximate the error rate in a validation data set. In other words, the training error rate approximates the prediction (extra-sample) error rate.

Related Techniques

NPMR is essentially a smoothing technique that can be cross-validated and applied in a predictive way. Many other smoothing techniques are well known, for example smoothing splines and wavelets. The optimal choice of a smoothing method depends on the specific application. Nonparametric regression models always fits for larger data.

Regression Trees

Decision tree learning algorithms can be applied to learn to predict a dependent variable from data. Although the original Classification And Regression Tree (CART) formulation applied only to predicting univariate data, the framework can be used to predict multivariate data including time series.

Semiparametric Regression

A semiparametric model is a regression model with both a finite- and an infinite-dimensional component.

A *finite-dimensional component* is spanned by some list of vectors (a vector is an object that has both magnitude and direction). The two-dimensional and three-dimensional spaces we deal with in everyday geometry are examples of finite-dimensional spaces, but so is a hypothetical 4568-dimensional space. *Infinite dimensional spaces* are spaces that have an infinite, and possibly ill-defined, number of dimensions and possibilities. They aren't spanned by any finite list of vectors.

In contrast to parametric models, which are well-defined in the finite-dimensional space, and non-parametric models, where the parameters can all span an infinite space, a semiparametric model has a component that is finite-dimensional (i.e. it's easy to research and understand), and another that is infinite-dimensional (i.e. beyond the range of ordinary statistical methods).

In research and statistics projects involving semi-parametric models the emphasis is almost always on the parametric component of the model. That's because this is the part which lends itself well to research.

Use of Semiparametric Models

Too often parametric models, while being easy to understand and easy to work with, fail to give a fair representation of what is happening in the real world. Non-parametric models may be better representations but do not lend themselves well to analysis. A semiparametric model allows you to have the best of both worlds: a model that is understandable and can be manipulated while still offering a fair representation of the messiness that is involved in real life.

Examples of Semiparametric Models

One example of a semi-parametric model is the Cox Proportional Hazards Model. This model is very useful in studies of the time remaining before an end or failure; it's used when doing research on the time remaining before a patient dies, or before a light bulb burns out. It's defined as:

$$F(t)=1-\exp(-\int_0^t \lambda_0(u)e^{\beta' x}du)$$

Here x is what we call the covariate vector, and our unknown parameters are B and $\lambda_0(u)$. B is finite-dimensional and is often the subject of research; $\lambda_0(u)$. on the other hand, is a unknown, non-negative function of time, and the space of $\lambda_0(u)$. possibilities is infinite. We call it a 'nuisance parameter'; a parameter that can't be entirely omitted in a rigorous treatment of a problem but is not of immediate interest– typically because it is difficult or impossible to study.

Methods

Many different semiparametric regression methods have been proposed and developed. The most popular methods are the partially linear, index and varying coefficient models.

Partially Linear Models

A partially linear model is given by

$$Y_i = X_i'\beta + g(Z_i) + u_i, \quad i=1,\ldots,n,$$

where Y_i is the dependent variable, X_i is a $p\times1$ vector of explanatory variables, β is a $p\times1$ vector of unknown parameters and $Z_i \in \mathrm{R}^q$. The parametric part of the partially linear model is given by the parameter vector β while the nonparametric part is the unknown function $g(Z_i)$. The data is assumed to be i.i.d. with $E(u_i \mid X_i, Z_i)=0$ and the model allows for a conditionally heteroskedastic error process $E(u_i^2 \mid x,z) = \sigma^2(x,z)$ of unknown form. This type of model was proposed by Robinson (1988) and extended to handle categorical covariates by Racine and Li.

This method is implemented by obtaining a $\sqrt{n}$ consistent estimator of β and then deriving an estimator of $g(Z_i)$ from the nonparametric regression of $Y_i - X_i'\hat{\beta}$ on z using an appropriate nonparametric regression method.

Index Models

A single index model takes the form

$$Y = g(X'\beta_0) + u,$$

where Y, X and β_0 are defined as earlier and the error term u satisfies $E(u \mid X) = 0$. The single index model takes its name from the parametric part of the model $x'\beta$ which is a *scalar* single index. The nonparametric part is the unknown function $g(\cdot)$.

Ichimura's Method

The single index model method developed by Ichimura is as follows. Consider the situation in which y is continuous. Given a known form for the function $g(\cdot)$, β_0 could be estimated using the nonlinear least squares method to minimize the function

$$\sum_{i=1} \left(Y_i - g(X_i'\beta)\right)^2.$$

Since the functional form of $g(\cdot)$ is not known, we need to estimate it. For a given value for β an estimate of the function

$$G(X_i'\beta) = E(Y_i \mid X_i'\beta) = E\left[g(X_i'\beta_0) \mid X_i'\beta\right]$$

using kernel method. Ichimura proposes estimating $g(X_i'\beta)$ with

$$\hat{G}_{-i}(X_i'\beta),$$

the leave-one-out nonparametric kernel estimator of $G(X_i'\beta)$.

Klein and Spady's Estimator

If the dependent variable y is binary and X_i and u_i are assumed to be independent, Klein and Spady propose a technique for estimating β using maximum likelihood methods. The log-likelihood function is given by:

$$L(\beta) = \sum_i (1 - Y_i)\ln\left(1 - \hat{g}_{-i}(X_i'\beta)\right) + \sum_i Y_i \ln\left(\hat{g}_{-i}(X_i'\beta)\right),$$

where $\hat{g}_{-i}(X_i'\beta)$ is the leave-one-out estimator.

Smooth Coefficient/Varying Coefficient Models

Hastie and Tibshirani propose a smooth coefficient model given by

$$Y_i = \alpha(Z_i) + X_i'\beta(Z_i) + u_i = (1 + X_i')\begin{pmatrix}\alpha(Z_i)\\ \beta(Z_i)\end{pmatrix} + u_i = W_i'\gamma(Z_i) + u_i,$$

Where X_i is a $k \times 1$ vector and $\beta(z)$ is a vector of unspecified smooth functions of z.

$\gamma(\cdot)$ may be expressed as

$$\gamma(Z_i) = \left(E[W_i W_i' \mid Z_i]\right)^{-1} E[W_i Y_i \mid Z_i].$$

Regression Models

Poisson Regression

To define the basic version of a Poisson regression model, suppose that we have observations $y_1,\ldots,y_n$ for the response variable $y_1,\ldots,y_n$ assumed to be independently distributed Poisson variates with means $\mu_1,\ldots,\mu_n$, i.e.

$$f(y_i|\mu_i) = \frac{\mu_i^{y_i}}{y_i!} = \exp(-\mu_i)$$

The systematic component of the model is specified by some regression function η, depending on regression parameters $\beta_1,\ldots,\beta_k$, with each component relating values $x_{i1},\ldots,x_{ik}$ of explanatory variables to respective means, i.e.

$$\mu_i = \eta_i(\beta) = \eta i(x_i 1,\ldots,x_{ik};\ \beta_1,\ldots,\beta_k).$$

Below Table Distribution by months prior to interview of stressful events reported from subjects: 147 subjects reporting exactly one stressful event in the period from 1 to 18 months prior to interview.

Months 1 2 3 4 5 6 7 8 9 10 11 12 13 14 15 16 17 18
Number 15 11 14 17 5 11 10 4 8 10 7 9 11 3 6 1 1 4

Often, this relationship is such that some monotone transformation g of the means is connected to a linear predictor of explanatory variables,

$$g(\mu_i) = \sum_{j=1}^{k} x_{ij}\beta_j.$$

In this situation g is called the link function and the model defined in this manner is an instance of a generalized linear model. For $\eta_i(\beta) = \left(\sum_{j=1}^{k} xi_j \beta_j\right)$ we have the familiar log linear model,

$$\log \mu_i = \sum_{j=1}^{k} x_{ij}\beta_j.$$

For the model specified by the stochastic component $f(y_i|\mu_i) = \frac{\mu_i^{y_i}}{y_i!} = \exp(-\mu_i)$ and regression function $\mu_i = \eta_i(\beta) = \eta i(x_i 1,\ldots,x_{ik};\ \beta_1,\ldots,\beta_k).$, the log likelihood function is written as:

$$\ell_y(\beta) = \sum_{i=1}^{n}\{y_i \log[\eta_i(\beta)] - \eta_i(\beta) - log(y_i!)\}.$$

It may be worthwhile noting that this reduces to a k-parameter exponential family log likelihood,

$$\ell_y(\beta) = \sum_{j=1}^{k}\left(\sum_{i=1}^{n} x_{ij}\, y_i\right)\beta_i - \sum_{i=1}^{n}\exp\left(\sum_{j=1}^{k} x_{ij}\beta_j\right)$$
$$-\sum_{i=1}^{n}\log(y_i!),$$

with jointly sufficient statistics $\sum_{i=1}^{n} x_{ij}\, y_i, j = 1,\ldots\ldots,k,$ if the model is log linear.

Some Special Cases

Loglinear Models for Contingency Tables Suppose, in obvious notation, y_{ij} with indices i = 1,...,I and j = 1,...,J form a two-dimensional contingency table, according to some classifying factors A and B having I and J categories, respectively. A common method for analyzing data of this kind is to assume that cell frequencies y_{ij} are independently distributed as Poisson and to use loglinear models, where in an analysis-of-variance-like fashion logarithms of expected cell frequencies μ_{ij} are assumed to be sums of several effects, e.g. for the multiplicative model,

$$\log(\mu_{ij}) = \beta_o + \beta_i^A + \beta_j^B,$$

subject to some constraints on the β_s. Sums of independent Poisson variates are again distributed as Poisson with means equal to the sum of respective means. Row totals Y_{i+}, column totals Y_{+j}, and grand total Y_{++} are, therefore, Poisson variates with means $\mu_{i+} = \mu_{i1} + \cdots + \mu_{iJ}$, $\mu_{+j} = \mu_{1j} + \cdots + \mu_{Ij}$, and $\mu_{++} = \sum_{i,j}\mu_{ij}$, respectively. Under the assumption of the multiplicative model these quantities are related by

$$\mu_{ij} = \frac{\mu_i + \mu + j}{\mu + +},$$

showing that the joint distribution of the contingency table is, in a multiplicative manner, completely determined by the marginal distributions.

The Poisson model assumption implies that marginals are random. If, instead, the total is fixed by the sampling design, it may be more appropriate to assume a multinomial distribution for the table. Formally, the multinomial model can be inferred from the Poisson model by conditioning on the total y_{++}. For the probability πij of an observation falling into row i and column j, we then have $\pi_{ij} = \mu_{ij} / \mu_{++}$, and from assuming the multiplicative model $\log(\mu_{ij}) = \beta_o + \beta_i^A + \beta_j^B$, it follows that

$$\pi_{ij} = \pi_i + \pi_{+j} ,$$

Where π_{i+} and π_{+j} are the marginal probabilities of an observation falling into row i and column j, respectively. Hence, row and column variables A and B are independently distributed.

Likewise, if row totals are fixed, then each row may be assumed to be multinomially distributed. Again, this can be inferred from the Poisson model by conditioning on the row totals, and the multiplicative model $\log(\mu_{ij}) = \beta_o + \beta_i^A + \beta_j^B$, implies identical distributions for the rows – a condition usually called homogeneity. It may be worthwhile noting that maximum-likelihood estimates for the parameters in the Poisson models are identical to those obtained for some other sampling designs, such as the multinomial designs just mentioned, making this class of model particularly interesting and useful.

Loglinear models for two- and higher-dimensional contingency tables, used to describe the association and interaction structure connecting the variables, are

Table: Number of recurrences of superficial bladder cancer for 31 male patients with grade 2, stage T1, solitary primary tumors and respective times under observation (in months) by size of primary tumor. Subset of data analyzed in,

Size	Recurrences	Time under observation
≤3cm	1	2, 3, 6, 8, 9, 10, 11, 13, 14, 16, 21, 22, 24, 26, 27
	2	7, 13, 15, 18, 23
	3	20
	4	24
>3cm	1	1, 5, 17, 18, 25
	2	18, 25
	3	4
	4	19

Usually, the goal is to find a parsimonious model that fits the data well and allows meaningful substantive interpretation. Most commonly, this search is restricted to hierarchical models.

Multiplicative Rate Models

If occurrences of some kind of event are counted over time, then often interest lies in the rate at which events occur. The rate describes the instantaneous risk for an event to happen at a given point in time. To be more specific, the probability of observing exactly one event in the interval ranging from t to t + h, divided by its length h, is assumed to tend to some value λ(t), as h tends to 0. λ(t), as a function of time t, is called the rate or intensity function.

An important special case, termed the Poisson process, assumes that waiting times between successive events are independent and exponentially distributed with common mean 1/λ. Here, the rate function is constant over time, λ(t) ≡ λ. Furthermore, the number Y (t) of events that occur up to time t is distributed as Poisson with mean $\mu = \lambda t$. Note that the mean of Y (t)/t equals the rate λ. This suggests a Poisson regression approach,

$$log\ \lambda = log\left(\frac{\mu}{t}\right) = \alpha + \beta x$$

for modeling the dependence of the rate function on an explanatory variable X. This can be rewritten as:

$$log\mu = \alpha + \beta x + log\ t,$$

with log(t) as an offset, i.e. a variable in the linear predictor, the corresponding regression parameter of which is set equal to 1. Observe that this defines a multiplicative model for the rate function,

$$\lambda = e^{\alpha}\left(e^{\beta}\right)^{x},$$

with a baseline rate $\lambda_o = \exp(\alpha)$ and proportionality factor $\exp(\beta x)$.

For illustrative purposes a subset of the data analyzed in is reprinted in above Table. For 31 male patients, who have been treated for superficial bladder cancer, the number of recurrent tumors has been recorded for some time after removal of the primary tumor. Defining X to be 1 for larger primary tumors (>3 cm) and 0 otherwise, and assuming a Poisson process with rate, yields parameter estimates $\hat{\alpha} = -1.95$ and $\hat{\beta} = 0.385$. The (baseline) rate for smaller tumors is (estimated as) 0.142, the rate for larger tumors being 1.47 times larger. In terms of waiting times between recurrences, means are estimated as 7.06 and 4.80 months, respectively.

Now suppose that we have recorded, for n individuals, time under observation, t_i, and the number y_i of events occurred. Observation times are assumed to be nonrandom and counts to be mutually independent. We also have a set of explanatory variables $x_{i1},\dots,x_{ik}$ available for each subject. Under the assumption of proportional rates, $\lambda_i = \lambda_o \exp\sum_{j=1}^{k} x_{ij}\beta_j$ we have,

$$\mu_i = \lambda_i \times t_i = \lambda_o \exp\left[\log(t_i) + \sum_{j=1}^{k} x_{ij}\beta_j\right]$$
$$= \lambda_o t_i \prod_{j=1}^{k} \exp\left(x_{ij}\beta_j\right),$$

i.e. a loglinear model for the mean of the Poisson process, involving the logarithm of observation times as an “explanatory” variable, with the associated regression parameter fixed at a value of 1.

If the process is such that it can be characterized by a time-varying rate function $\lambda(t)$, it is called a nonhomogeneous Poisson process. Writing

$$\Lambda(t) = \int_0^t \lambda(u)\,du$$

for the integrated rate or intensity function, the number of occurrences of the event in period until time point t is again distributed as Poisson, but with mean equal to $\Lambda(t)$. Note that events in nonoverlapping time intervals are independent, but waiting times between successive events are, contrary to the homogeneous process with constant rate, neither identically distributed nor independent. In this situation model,

$$\mu_i = \lambda_i \times t_i = \lambda_0 \exp\left[\log(t_i) + \sum_{j=1}^{k} x_{ij}\beta_j\right] = \lambda_0 t_i \prod_{j=1}^{k} \exp\left(x_{ij}\beta_j\right),$$

can be modified, using a baseline rate function $\lambda_0\ (t|\alpha)$, possibly depending on some additional parameter α, to give,

$$\mu_i = \exp\left\{ log\left[\Lambda_0\left(t_i \mid \alpha\right)\right] + \sum_{j=1}^{k} x_{ij}\beta_j \right\}.$$

Choosing $\Lambda_0(t|\alpha)$ to be t,tα, or exp(αt) corresponds to an exponential, a Weibull, or an extreme value intensity function, respectively, and results in a loglinear model for the Y_{is}. Disregarding constant terms, the likelihood function for this model is

$$L(\alpha,\beta) = \prod_{i=1}^{n} \left[\Lambda_0\left(t_i \mid \alpha\right) \exp\left(\sum_{j=1}^{k} x_{ij}\beta_j\right)\right]^{y_i} \times \exp\left[-\Lambda_0\left(t_i \mid \alpha\right) \exp\left(\sum_{j=1}^{k} x_{ij}\beta_j\right)\right].$$

If times for occurrences of each event were known, a multiplicative term, depending on the parameter α, would be added to above euation.

There is a close connection to relative risk models, which are very frequently used in epidemiology. This class of models assumes that risk factors interact in a multiplicative way.

Maximum Likelihood-based Parameter Estimation

Given a set of parameters θ and an input vector x, the mean of the predicted Poisson distribution, as stated above, is given by,

$$\lambda = \mathrm{E}(Y \mid x) = e^{\theta' x}$$

and thus, the Poisson distribution's probability mass function is given by,

$$p(y \mid x;\theta) = \frac{\lambda^y}{y!} e^{-\lambda} = \frac{e^{y\theta' x} e^{-e^{\theta' x}}}{y!}.$$

Now suppose we are given a data set consisting of m vectors $x_i \in \mathbb{R}^{n+1}, i = 1,\ldots,m$, along with a set of m values $y_1,\ldots,y_m \in \mathbb{N}$. Then, for a given set of parameters θ, the probability of attaining this particular set of data is given by,

$$p(y_1,\ldots,y_m \mid x_1,\ldots,x_m;\theta) = \prod_{i=1}^{m} \frac{e^{y_i\theta' x_i} e^{-e^{\theta' x_i}}}{y_i!}.$$

By the method of maximum likelihood, we wish to find the set of parameters θ that makes this probability as large as possible. To do this, the equation is first rewritten as a likelihood function in terms of θ:

$$L(\theta \mid X,Y) = \prod_{i=1}^{m} \frac{e^{y_i \theta' x_i} e^{-e^{\theta' x_i}}}{y_i!}.$$

Note that the expression on the right hand side has not actually changed. A formula in this form is typically difficult to work with; instead, one uses the *log-likelihood*:

$$\ell(\theta \mid X,Y) = \log L(\theta \mid X,Y) = \sum_{i=1}^{m} \left(y_i \theta' x_i - e^{\theta' x_i} - \log(y_i!) \right).$$

Notice that the parameters θ only appear in the first two terms of each term in the summation. Therefore, given that we are only interested in finding the best value for θ we may drop the $y_i!$ and simply write,

$$\ell(\theta \mid X,Y) = \sum_{i=1}^{m} \left(y_i \theta' x_i - e^{\theta' x_i} \right).$$

To find a maximum, we need to solve an equation $\frac{\partial \ell(\theta \mid X,Y)}{\partial \theta} = 0$ which has no closed-form solution. However, the negative log-likelihood, $-\ell(\theta \mid X,Y)$, is a convex function, and so standard convex optimization techniques such as gradient descent can be applied to find the optimal value of θ.

Poisson Regression in Practice

Poisson regression may be appropriate when the dependent variable is a count, for instance of events such as the arrival of a telephone call at a call centre. The events must be independent in the sense that the arrival of one call will not make another more or less likely, but the probability per unit time of events is understood to be related to covariates such as time of day.

"Exposure" and Offset

Poisson regression may also be appropriate for rate data, where the rate is a count of events divided by some measure of that unit's *exposure* (a particular unit of observation). For example, biologists may count the number of tree species in a forest: events would be tree observations, exposure would be unit area, and rate would be the number of species per unit area. Demographers may model death rates in geographic areas as the count of deaths divided by person−years. More generally, event rates can be calculated as events per unit time, which allows the observation window to vary for each unit. In these examples, exposure is respectively unit area, person−years and unit time. In Poisson regression this is handled as an offset, where the exposure variable enters on the right-hand side of the equation, but with a parameter estimate (for log(exposure)) constrained to 1.

$$\log(\mathrm{E}(Y \mid x)) = \log(\text{exposure}) + \theta' x$$

which implies

$$\log(\mathrm{E}(Y \mid x)) - \log(\text{exposure}) = \log\left(\frac{\mathrm{E}(Y \mid x)}{\text{exposure}}\right) = \theta' x$$

Offset in the case of a GLM in R can be achieved using the offset() function:

```
glm(y ~ offset(log(exposure)) + x, family=poisson(link=log) )
```

Overdispersion and Zero Inflation

A characteristic of the Poisson distribution is that its mean is equal to its variance. In certain circumstances, it will be found that the observed variance is greater than the mean; this is known as overdispersion and indicates that the model is not appropriate. A common reason is the omission of relevant explanatory variables, or dependent observations. Under some circumstances, the problem of overdispersion can be solved by using quasi-likelihood estimation or a negative binomial distribution instead.

Ver Hoef and Boveng described the difference between quasi-Poisson (also called overdispersion with quasi-likelihood) and negative binomial (equivalent to gamma-Poisson) as follows: If $E(Y) = \mu$, the quasi-Poisson model assumes $\mathrm{var}(Y) = \theta\mu$ while the gamma-Poisson assumes $\mathrm{var}(Y) = \mu(1 + \kappa\mu)$, where θ is the quasi-Poisson overdispersion parameter, and κ is the shape parameter of the negative binomial distribution. For both models, parameters are estimated using Iteratively reweighted least squares. For quasi-Poisson, the weights are μ/θ. For negative binomial, the weights are $\mu/(1 + \kappa\mu)$. With large μ and substantial extra-Poisson variation, the negative binomial weights are capped at $1/\kappa$. Ver Hoef and Boveng discussed an example where they selected between the two by plotting mean squared residuals vs. the mean.

Another common problem with Poisson regression is excess zeros: if there are two processes at work, one determining whether there are zero events or any events, and a Poisson process determining how many events there are, there will be more zeros than a Poisson regression would predict. An example would be the distribution of cigarettes smoked in an hour by members of a group where some individuals are non-smokers.

Other generalized linear models such as the negative binomial model or zero-inflated model may function better in these cases.

Use in Survival Analysis

Poisson regression creates proportional hazards models, one class of survival analysis.

Ordinal Regression

In statistics, ordinal regression (also called "ordinal classification") is a type of regression analysis used for predicting an ordinal variable, i.e. a variable whose value exists on an arbitrary scale where only the relative ordering between different values is significant. It can be considered an intermediate problem between regression and classification. Examples of ordinal regression are ordered logit and ordered probit. Ordinal regression turns up often in the social sciences, for example in the modeling of human levels of preference (on a scale from, say, 1–5 for "very poor" through "excellent"), as well as in information retrieval. In machine learning, ordinal regression may also be called ranking learning.

Linear Models for Ordinal Regression

Ordinal regression can be performed using a generalized linear model (GLM) that fits both a coefficient vector and a set of *thresholds* to a dataset. Suppose one has a set of observations, represented by length-p vectors x_1 through x_n, with associated responses y_1 through y_n, where each y_i is an ordinal variable on a scale 1, ..., K. To this data, one fits a length-p coefficient vector w and a set of thresholds $\theta_1, ..., \theta_{K-1}$ with the property that $\theta_1 < \theta_2 < ... < \theta_{K-1}$. This set of thresholds divides the real number line into K disjoint segments, corresponding to the K response levels.

The model can now be formulated as:

$$\Pr(y \leq i \mid \mathbf{x}) = \sigma(\theta_i - \mathbf{w} \cdot \mathbf{x})$$

or, the cumulative probability of the response y being at most i is given by a function σ (the inverse link function) applied to a linear function of x. Several choices exist for σ; the logistic function,

$$\sigma(\theta_i - \mathbf{w} \cdot \mathbf{x}) = \frac{1}{1 + e^{-(\theta_i - \mathbf{w} \cdot \mathbf{x})}}$$

gives the ordered logit model, while using the probit function gives the ordered probit model. A third option is to use an exponential function,

$$\sigma(\theta_i - \mathbf{w} \cdot \mathbf{x}) = \exp(-\exp(\theta_i - \mathbf{w} \cdot \mathbf{x}))$$

which gives the proportional hazards model.

Latent Variable Model

The probit version of the above model can be justified by assuming the existence of a real-valued latent variable (unobserved quantity) y^*, determined by:

$$y^* = \mathbf{w} \cdot \mathbf{x} + \varepsilon$$

where ε is normally distributed with zero mean and unit variance, conditioned on x. The response variable y results from an "incomplete measurement" of y^*, where one only determines the interval into which y^* falls:

$$y = \begin{cases} 1 \text{ if } y^* \leq \theta_1, \\ 2 \text{ if } \theta_1 < y^* \leq \theta_2, \\ 3 \text{ if } \theta_2 < y^* \leq \theta_3 \\ \vdots \\ K \text{ if } \theta_{K-1} < y^*. \end{cases}$$

Defining $\theta_0 = -\infty$ and $\theta_K = \infty$, the above can be summarized as $y = k$ if and only if $\theta_{k-1} < y^* \leq \theta_k$.

From these assumptions, one can derive the conditional distribution of y as:

$$\begin{aligned} P(y=k\,|\,\mathbf{x}) &= P(\theta_{k-1} < y^* \leq \theta_k\,|\,\mathbf{x}) \\ &= P(\theta_{k-1} < \mathbf{w}\cdot\mathbf{x}+\varepsilon \leq \theta_k) \\ &= \Phi(\theta_k - \mathbf{w}\cdot\mathbf{x}) - \Phi(\theta_{k-1} - \mathbf{w}\cdot\mathbf{x}) \end{aligned}$$

where Φ is the cumulative distribution function of the standard normal distribution, and takes on the role of the inverse link function σ. The log-likelihood of the model for a single training example x_i, y_i can now be stated as:

$$\log L\left(w,\theta\middle|x_i,y_i\right) = \sum_{k=1}^{k}[y_i = k]\log\left[\Phi\left(\theta_k - w.x_i\right) - \Phi\left(\theta_{k-1} - w.x_i\right)\right]$$

(using the Iverson bracket $[y_i = k]$.) The log-likelihood of the ordered logit model is analogous, using the logistic function instead of Φ.

Alternative Models

In machine learning, alternatives to the latent-variable models of ordinal regression have been proposed. An early result was PRank, a variant of the perceptron algorithm that found multiple parallel hyperplanes separating the various ranks; its output is a weight vector w and a sorted vector of $K-1$ thresholds θ, as in the ordered logit/probit models. The prediction rule for this model is to output the smallest rank k such that $wx < \theta_k$.

Other methods rely on the principle of large-margin learning that also underlies support vector machines.

Another approach is given by Rennie and Srebro, who, realizing that "even just evaluating the likelihood of a predictor is not straight-forward" in the ordered logit and ordered probit models, propose fitting ordinal regression models by adapting common loss functions from classification (such as the hinge loss and log loss) to the ordinal case.

Software

ORCA (Ordinal Regression and Classification Algorithms) is a Octave/MATLAB framework including a wide set of ordinal regression methods.

Ordinary Least Squares (OLS) Regression Model

In statistics, ordinary least squares (OLS) is a type of linear least squares method for estimating the unknown parameters in a linear regression model. OLS chooses the parameters of a linear function of a set of explanatory variables by the principle of least squares: minimizing the sum of the squares of the differences between the observed dependent variable (values of the variable being predicted) in the given dataset and those predicted by the linear function.

Geometrically, this is seen as the sum of the squared distances, parallel to the axis of the dependent variable, between each data point in the set and the corresponding point on the regression surface – the smaller the differences, the better the model fits the data. The resulting estimator can be expressed by a simple formula, especially in the case of a simple linear regression, in which there is a single regressor on the right side of the regression equation.

The OLS estimator is consistent when the regressors are exogenous, and optimal in the class of linear unbiased estimators when the errors are homoscedastic and serially uncorrelated. Under these conditions, the method of OLS provides minimum-variance mean-unbiased estimation when the errors have finite variances. Under the additional assumption that the errors are normally distributed, OLS is the maximum likelihood estimator.

OLS is used in fields as diverse as economics (econometrics), political science, psychology and engineering (control theory and signal processing).

Linear Model

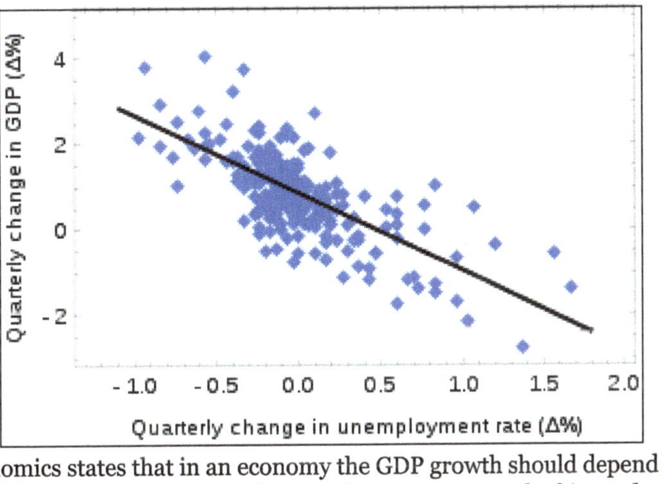

Okun's law in macroeconomics states that in an economy the GDP growth should depend linearly on the changes in the unemployment rate. Here the ordinary least squares method is used to construct the regression line describing this law.

Suppose the data consist of n observations $\{y_i, x_i\}^n_{i=1}$. Each observation i includes a scalar response y_i and a column vector x_i of values of p predictors (regressors) x_{ij} for $j = 1, ..., p$. In a linear regression model, the response variable, y_i, is a linear function of the regressors:

$$y_i = \beta_1 x_{i1} + \beta_2 x_{i2} + \cdots + \beta_p x_{ip} + \varepsilon_i,$$

or in vector form,

$$y_i = x_i^T \beta + \varepsilon_i,$$

where β is a $p \times 1$ vector of unknown parameters; the ε_i's are unobserved scalar random variables (errors) which account for influences upon the responses y_i from sources other than the explanators x_i; and x_i is a column vector of the ith observations of all the explanatory variables. This model can also be written in matrix notation as:

$$y = X\beta + \varepsilon,$$

where y and ε are $n \times 1$ vectors of the values of the response variable and the errors for the various observations, and X is an $n \times p$ matrix of regressors, also sometimes called the design matrix, whose row i is x_i^T and contains the ith observations on all the explanatory variables.

As a rule, the constant term is always included in the set of regressors X, say, by taking $x_{i1} = 1$ for all $i = 1, \ldots, n$. The coefficient β_1 corresponding to this regressor is called the *intercept*.

There may be some relationship between the regressors. For instance, the third regressor may be the square of the second regressor. In this case (assuming that the first regressor is constant) we have a quadratic model in the second regressor. But this is still considered a linear model because it is linear in the βs.

Matrix/vector Formulation

Consider an overdetermined system of m linear equations in n unknown coefficients, $\beta_1, \beta_2, \ldots, \beta_n$, with $m > n$.

$$\sum_{j=1}^{n} X_{ij}\beta_j = y_i, (i = 1, 2, \ldots, m),$$

For a linear model as above, not all of X contains information on the data points. The first column is populated with ones, $X_{i1} = 1$, only the other columns contain actual data, and n = number of regressors + 1. This can be written in matrix form as:

$$\mathbf{X}\boldsymbol{\beta} = \mathbf{y},$$

where;

$$\mathbf{X} = \begin{bmatrix} X_{11} & X_{12} & \cdots & X_{1n} \\ X_{21} & X_{22} & \cdots & X_{2n} \\ \vdots & \vdots & \ddots & \vdots \\ X_{m1} & X_{m2} & \cdots & X_{mn} \end{bmatrix}, \quad \boldsymbol{\beta} = \begin{bmatrix} \beta_1 \\ \beta_2 \\ \vdots \\ \beta_n \end{bmatrix}, \quad \mathbf{y} = \begin{bmatrix} y_1 \\ y_2 \\ \vdots \\ y_m \end{bmatrix}.$$

Such a system usually has no solution, so the goal is instead to find the coefficients $\hat{\mathbf{a}}$ which fit the equations "best", in the sense of solving the quadratic minimization problem.

$$\hat{\boldsymbol{\beta}} = \underset{\boldsymbol{\beta}}{\arg\min}\, S(\boldsymbol{\beta}),$$

where the objective function S is given by:

$$S(\boldsymbol{\beta}) = \sum_{i=1}^{m} \left| y_i - \sum_{j=1}^{n} X_{ij}\beta_j \right|^2 = \left\| \mathbf{y} - \mathbf{X}\boldsymbol{\beta} \right\|^2.$$

A justification for choosing this criterion is given in properties below. This minimization problem has a unique solution, provided that the n columns of the matrix $\mathbf{X}$ are linearly independent, given by solving the normal equations.

$$(X^TX)\hat{\beta} = X^Ty.$$

The matrix $\mathbf{X}^T\mathbf{X}$ is known as the Gramian matrix of $\mathbf{X}$, which possesses several nice properties such as being a positive semi-definite matrix, and the matrix $\mathbf{X}^T\mathbf{y}$ is known as the moment matrix of regressand by regressors. Finally, $\hat{\beta}$ is the coefficient vector of the least-squares hyperplane, expressed as:

$$\hat{\beta} = (X^TX)^{-1}X^Ty.$$

Estimation

Suppose b is a "candidate" value for the parameter vector β. The quantity $y_i - x_i^Tb$, called the residual for the i-th observation, measures the vertical distance between the data point $(x_i\, y_i)$ and the hyperplane $y = x^Tb$, and thus assesses the degree of fit between the actual data and the model. The sum of squared residuals (SSR) (also called the error sum of squares (ESS) or residual sum of squares (RSS)) is a measure of the overall model fit:

$$S(b) = \sum_{i=1}^{n}(y_i - x_i^Tb)^2 = (y - Xb)^T(y - Xb),$$

where T denotes the matrix transpose, and the rows of X, denoting the values of all the independent variables associated with a particular value of the dependent variable, are $X_i = x_i^T$. The value of b which minimizes this sum is called the OLS estimator for $\boldsymbol{\beta}$. The function $S(b)$ is quadratic in b with positive-definite Hessian, and thercfore this function possesses a unique global minimum at $b = \hat{\beta},$, which can be given by the explicit formula:

$$\hat{\beta} = \operatorname{argmin}_{b\in\mathbb{R}^p} S(b) = (X^TX)^{-1}X^Ty\,.$$

The product $N=X^T X$ is a normal matrix and its inverse, $Q=N^{-1}$, is the *cofactor matrix* of β, closely related to its covariance matrix, C_β. The matrix $(X^T X)^{-1} X^T=Q X^T$ is called the Moore–Penrose pseudoinverse matrix of X. This formulation highlights the point that estimation can be carried out if, and only if, there is no perfect multicollinearity between the explanatory variables (which would cause the normal matrix to have no inverse).

After we have estimated β, the fitted values (or predicted values) from the regression will be:

$$\hat{y} = X\hat{\beta} = Py,$$

where $P = X(X^TX)^{-1}X^T$ is the projection matrix onto the space V spanned by the columns of X. This matrix P is also sometimes called the hat matrix because it "puts a hat" onto the variable y. Another matrix, closely related to P is the *annihilator* matrix $M = I_n - P$; this is a projection matrix onto the space orthogonal to V. Both matrices P and M are symmetric and idempotent (meaning that $P^2 = P$), and relate to the data matrix X via identities $PX = X$ and $MX = 0$. Matrix M creates the residuals from the regression:

$$\hat{\varepsilon} = y - \hat{y} = y - X\hat{\beta} = My = M(X\beta + \varepsilon) = (MX)\beta + M\varepsilon = M\varepsilon.$$

Using these residuals we can estimate the value of σ^2, called the reduced chi-squared:

$$s^2 = \frac{\hat{\varepsilon}^{\mathrm{T}}\hat{\varepsilon}}{n-p} = \frac{(My)^{\mathrm{T}} My}{n-p} = \frac{y^{\mathrm{T}} M^{\mathrm{T}} My}{n-p} = \frac{y^{\mathrm{T}} My}{n-p} = \frac{S(\hat{\beta})}{n-p}, \qquad \hat{\sigma}^2 = \frac{n-p}{n} s^2$$

The numerator, $n-p$, is the statistical degrees of freedom. The first quantity, s^2, is the OLS estimate for σ^2, whereas the second, $\hat{\sigma}^2$, is the MLE estimate for σ^2. The two estimators are quite similar in large samples; the first one is always unbiased, while the second is biased but minimizes the mean squared error of the estimator. In practice s^2 is used more often, since it is more convenient for the hypothesis testing. The square root of s^2 is called the regression standard error, standard error of the regression, or standard error of the equation.

It is common to assess the goodness-of-fit of the OLS regression by comparing how much the initial variation in the sample can be reduced by regressing onto X. The coefficient of determination R^2 is defined as a ratio of "explained" variance to the "total" variance of the dependent variable y:

$$R^2 = \frac{\sum(\hat{y}_i - \bar{y})^2}{\sum(y_i - \bar{y})^2} = \frac{y^{\mathrm{T}} P^{\mathrm{T}} LPy}{y^{\mathrm{T}} Ly} = 1 - \frac{y^{\mathrm{T}} My}{y^{\mathrm{T}} Ly} = 1 - \frac{\mathrm{RSS}}{\mathrm{TSS}}$$

where TSS is the total sum of squares for the dependent variable, $L = I_n - \mathbf{1}\mathbf{1}^T / n$, and $\mathbf{1}$ is an $n \times 1$ vector of ones. (L is a "centering matrix" which is equivalent to regression on a constant; it simply subtracts the mean from a variable.) In order for R^2 to be meaningful, the matrix X of data on regressors must contain a column vector of ones to represent the constant whose coefficient is the regression intercept. In that case, R^2 will always be a number between 0 and 1, with values close to 1 indicating a good degree of fit.

The variance in the prediction of the independent variable as a function of the dependent variable is given in the Polynomial least squares.

Simple Linear Regression Model

If the data matrix X contains only two variables, a constant and a scalar regressor x_i, then this is called the "simple regression model". This case is often considered in the beginner statistics classes, as it provides much simpler formulas even suitable for manual calculation. The parameters are commonly denoted as (α, β):

$$y_i = \alpha + \beta x_i + \varepsilon_i.$$

The least squares estimates in this case are given by simple formulas;

$$\hat{\beta} = \frac{\sum x_i y_i - \frac{1}{n}\sum x_i \sum y_i}{\sum x_i^2 - \frac{1}{n}(\sum x_i)^2} = \frac{\mathrm{Cov}[x,y]}{\mathrm{Var}[x]}$$

$$\hat{\alpha} = \bar{y} - \hat{\beta}\bar{x},$$

where Var(.) and Cov(.) are sample parameters.

Alternative Derivations

The least squares estimator $\hat{\beta}$ was obtained as a value that minimizes the sum of squared residuals of the model. However it is also possible to derive the same estimator from other approaches. In all cases the formula for OLS estimator remains the same: $\hat{\beta} = (X^T X)^{-1} X^T y$; the only difference is in how we interpret this result.

Projection

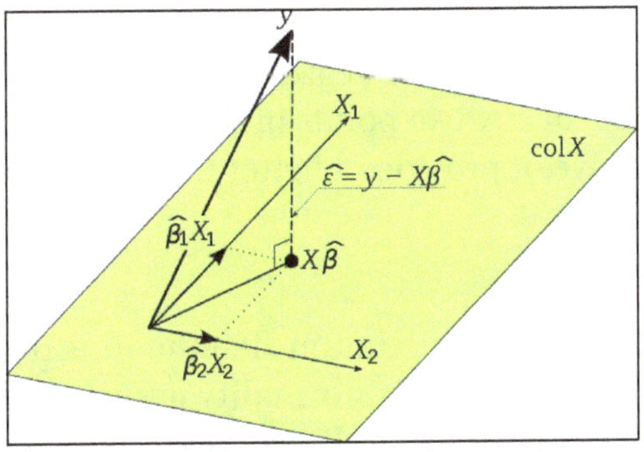

OLS estimation can be viewed as a projection onto the linear space spanned by the regressors. (Here each of X_1 and X_2 refers to a column of the data matrix.)

For mathematicians, OLS is an approximate solution to an over determined system of linear equations $X\beta \approx y$, where β is the unknown. Assuming the system cannot be solved exactly (the number of equations n is much larger than the number of unknowns p), we are looking for a solution that could provide the smallest discrepancy between the right- and left- hand sides. In other words, we are looking for the solution that satisfies,

$$\hat{\beta} = \arg\min_{\beta} \| y - X\beta \|,$$

where $\| \|$ is the standard L^2 norm in the n-dimensional Euclidean space R^n. The predicted quantity $X\beta$ is just a certain linear combination of the vectors of regressors. Thus, the residual vector $y - X\beta$ will have the smallest length when y is projected orthogonally onto the linear subspace spanned by the columns of X. The OLS estimator $\hat{\beta}$ in this case can be interpreted as the coefficients of vector decomposition of $\hat{y} = Py$ along the basis of X.

In other words, the gradient equations at the minimum can be written as:

$$(\mathbf{y} - X\hat{\boldsymbol{\beta}})^{\mathrm{T}} X = 0.$$

A geometrical interpretation of these equations is that the vector of residuals, $\mathbf{y} - X\hat{\boldsymbol{\beta}}$ is orthogonal to the column space of X, since the dot product $(\mathbf{y} - X\hat{\boldsymbol{\beta}}) \cdot X\mathbf{v}$ is equal to zero for *any* conformal vector, v. This means that $\mathbf{y} - X\hat{\boldsymbol{\beta}}$ is the shortest of all possible vectors $\mathbf{y} - X\boldsymbol{\beta}$, that is, the variance of the residuals is the minimum possible. This is illustrated at the right.

Introducing $\hat{\gamma}$ and a matrix K with the assumption that a matrix $[X\ K]$ is non-singular and $K^{\mathrm{T}} X = 0$ (cf. Orthogonal projections), the residual vector should satisfy the following equation:

$$\hat{\mathbf{r}} \triangleq \mathbf{y} - X\hat{\boldsymbol{\beta}} = K\hat{\gamma}.$$

The equation and solution of linear least squares are thus described as follows:

$$\mathbf{y} = [X \quad K]\begin{pmatrix}\hat{\beta}\\ \hat{\gamma}\end{pmatrix},$$

$$\begin{pmatrix}\hat{\beta}\\ \hat{\gamma}\end{pmatrix} = [X \quad K]^{-1}\mathbf{y} = \begin{bmatrix}(X^{\mathrm{T}}X)^{-1}X^{\mathrm{T}}\\ (K^{\mathrm{T}}K)^{-1}K^{\mathrm{T}}\end{bmatrix}\mathbf{y}.$$

Another way of looking at it is to consider the regression line to be a weighted average of the lines passing through the combination of any two points in the dataset. Although this way of calculation is more computationally expensive, it provides a better intuition on OLS.

Maximum Likelihood

The OLS estimator is identical to the maximum likelihood estimator (MLE) under the normality assumption for the error terms. This normality assumption has historical importance, as it provided the basis for the early work in linear regression analysis by Yule and Pearson. From the properties of MLE, we can infer that the OLS estimator is asymptotically efficient (in the sense of attaining the Cramér–Rao bound for variance) if the normality assumption is satisfied.

Generalized Method of Moments

In iid case the OLS estimator can also be viewed as a GMM estimator arising from the moment conditions,

$$\mathrm{E}\big[x_i(y_i - x_i^T\beta)\big] = 0.$$

These moment conditions state that the regressors should be uncorrelated with the errors. Since x_i is a p-vector, the number of moment conditions is equal to the dimension of the parameter vector β, and thus the system is exactly identified. This is the so-called classical GMM case, when the estimator does not depend on the choice of the weighting matrix.

Note that the original strict exogeneity assumption $E[\varepsilon_i | x_i] = 0$ implies a far richer set of moment conditions than stated above. In particular, this assumption implies that for any vector-function f, the moment condition $\mathrm{E}[f(x_i)\cdot\varepsilon_i] = 0$ will hold. However it can be shown using the Gauss–Markov theorem that the optimal choice of function f is to take $f(x) = x$, which results in the moment equation posted above.

Properties

Assumptions

There are several different frameworks in which the linear regression model can be cast in order to make the OLS technique applicable. Each of these settings produces the same formulas and same results. The only difference is the interpretation and the assumptions which have to be imposed in order for the method to give meaningful results. The choice of the applicable framework depends mostly on the nature of data in hand, and on the inference task which has to be performed.

One of the lines of difference in interpretation is whether to treat the regressors as random variables, or as predefined constants. In the first case (random design) the regressors x_i are random and sampled together with the y_i's from some population, as in an observational study. This approach allows for more natural study of the asymptotic properties of the estimators. In the other interpretation (fixed design), the regressors X are treated as known constants set by a design, and y is sampled conditionally on the values of X as in an experiment. For practical purposes, this distinction is often unimportant, since estimation and inference is carried out while conditioning on X.

Classical Linear Regression Model

The classical model focuses on the "finite sample" estimation and inference, meaning that the number of observations n is fixed. This contrasts with the other approaches, which study the asymptotic behavior of OLS, and in which the number of observations is allowed to grow to infinity.

- Correct specification. The linear functional form must coincide with the form of the actual data-generating process.
- Strict exogeneity. The errors in the regression should have conditional mean zero:

  $$\mathrm{E}[\varepsilon \mid X] = 0.$$

  The immediate consequence of the exogeneity assumption is that the errors have mean zero: $\mathrm{E}[\varepsilon] = 0$, and that the regressors are uncorrelated with the errors: $E\left[X^T\varepsilon\right] = 0$.

  The exogeneity assumption is critical for the OLS theory. If it holds then the regressor variables are called *exogenous*. If it doesn't, then those regressors that are correlated with the error term are called *endogenous*, and then the OLS estimates become invalid. In such case the method of instrumental variables may be used to carry out inference.

- No linear dependence. The regressors in X must all be linearly independent. Mathematically, this means that the matrix X must have full column rank almost surely:

  $$\Pr\left[\operatorname{rank}(X) = p\right] = 1.$$

  Usually, it is also assumed that the regressors have finite moments up to at least the second moment. Then the matrix $Q_{xx} = E\left[X^T X / n\right]$ is finite and positive semi-definite.

  When this assumption is violated the regressors are called linearly dependent or perfectly multicollinear. In such case the value of the regression coefficient β cannot be learned, although prediction of y values is still possible for new values of the regressors that lie in the same linearly dependent subspace.

- Spherical errors:

  $$\mathrm{Var}[\varepsilon \mid X] = \sigma^2 I_n,$$

where I_n is the identity matrix in dimension n, and σ^2 is a parameter which determines the variance of each observation. This σ^2 is considered a nuisance parameter in the model, although usually it is also estimated. If this assumption is violated then the OLS estimates are still valid, but no longer efficient.

It is customary to split this assumption into two parts:

- Homoscedasticity: $E\left[\varepsilon_i^2 | X\right] = \sigma^2$, which means that the error term has the same variance σ^2 in each observation. When this requirement is violated this is called heteroscedasticity, in such case a more efficient estimator would be weighted least squares. If the errors have infinite variance then the OLS estimates will also have infinite variance (although by the law of large numbers they will nonetheless tend toward the true values so long as the errors have zero mean). In this case, robust estimation techniques are recommended.
- No autocorrelation: the errors are uncorrelated between observations: $E\left[\varepsilon_i\varepsilon_j | X\right] = 0 \, fori \neq j$. This assumption may be violated in the context of time series data, panel data, cluster samples, hierarchical data, repeated measures data, longitudinal data, and other data with dependencies. In such cases generalized least squares provides a better alternative than the OLS. Another expression for autocorrelation is *serial correlation.*
- Normality. It is sometimes additionally assumed that the errors have normal distribution conditional on the regressors:

  $$\varepsilon | X \sim \mathcal{N}(0, \sigma^2 I_n).$$

  This assumption is not needed for the validity of the OLS method, although certain additional finite-sample properties can be established in case when it does (especially in the area of hypotheses testing). Also when the errors are normal, the OLS estimator is equivalent to the maximum likelihood estimator (MLE), and therefore it is asymptotically efficient in the class of all regular estimators. Importantly, the normality assumption applies only to the error terms; contrary to a popular misconception, the response (dependent) variable is not required to be normally distributed.

Independent and Identically Distributed (IID)

In some applications, especially with cross-sectional data, an additional assumption is imposed — that all observations are independent and identically distributed. This means that all observations are taken from a random sample which makes all the assumptions listed earlier simpler and easier to interpret. Also this framework allows one to state asymptotic results (as the sample size $n \to \infty$), which are understood as a theoretical possibility of fetching new independent observations from the data generating process. The list of assumptions in this case is:

- Id observations: (x_i, y_i) is independent from, and has the same distribution as, (x_j, y_j) for all $i \neq j$;
- No perfect multicollinearity: $Q_{xx} = \mathrm{E}[x_i x_i^{\mathrm{T}}]$ is a positive-definite matrix;
- Exogeneity: $\mathrm{E}[\varepsilon_i | x_i] = 0$;
- Homoscedasticity: $\mathrm{Var}[\varepsilon_i | x_i] = \sigma^2$.

Time Series Model

- The stochastic process $\{x_i, y_i\}$ is stationary and ergodic; if $\{x_i, y_i\}$ is nonstantionary, OLS results are often spurious unless $\{x_i, y_i\}$ is co-integrating.
- The regressors are *predetermined*: $E[x_i\varepsilon_i] = 0$ for all $i = 1, \ldots, n$;
- The $p \times p$ matrix $Q_{xx} = E[x_i x_i^T]$ is of full rank, and hence positive-definite;
- $\{x_i\varepsilon_i\}$ is a martingale difference sequence, with a finite matrix of second moments $Q_{xx\varepsilon^2} = E[\varepsilon_i^2 x_i x_i^T]$.

Finite Sample Properties

First of all, under the *strict exogeneity* assumption the OLS estimators $\hat{\beta}$ and s^2 are unbiased, meaning that their expected values coincide with the true values of the parameters:

$$E[\hat{\beta} \mid X] = \beta, \quad E[s^2 \mid X] = \sigma^2.$$

If the strict exogeneity does not hold (as is the case with many time series models, where exogeneity is assumed only with respect to the past shocks but not the future ones), then these estimators will be biased in finite samples.

The *variance-covariance matrix* (or simply *covariance matrix*) of $\hat{\beta}$ is equal to,

$$\operatorname{Var}[\hat{\beta} \mid X] = \sigma^2 (X^T X)^{-1} = \sigma^2 Q.$$

In particular, the standard error of each coefficient $\hat{\beta}_j$ is equal to square root of the j-th diagonal element of this matrix. The estimate of this standard error is obtained by replacing the unknown quantity σ^2 with its estimate s^2. Thus,

$$\widehat{\text{s.e.}}(\hat{\beta}_j) = \sqrt{s^2 (X^T X)^{-1}_{jj}}$$

It can also be easily shown that the estimator $\hat{\beta}$ is uncorrelated with the residuals from the model:

$$\operatorname{Cov}[\hat{\beta}, \hat{\varepsilon} \mid X] = 0.$$

The Gauss–Markov theorem states that under the *spherical errors* assumption (that is, the errors should be uncorrelated and homoscedastic) the estimator $\hat{\beta}$ is efficient in the class of linear unbiased estimators. This is called the best linear unbiased estimator (BLUE). Efficiency should be understood as if we were to find some other estimator $\hat{\beta}$ which would be linear in y and unbiased, then,

$$\operatorname{Var}[\tilde{\beta} \mid X] - \operatorname{Var}[\hat{\beta} \mid X] \geq 0$$

In the sense that this is a nonnegative-definite matrix. This theorem establishes optimality only in the class of linear unbiased estimators, which is quite restrictive. Depending on the distribution of the error terms ε, other, non-linear estimators may provide better results than OLS.

Assuming Normality

The properties listed so far are all valid regardless of the underlying distribution of the error terms. However, if you are willing to assume that the *normality assumption* holds (that is, that $\varepsilon \sim N(0, \sigma^2 I_n)$), then additional properties of the OLS estimators can be stated.

The estimator $\hat{\beta}$ is normally distributed, with mean and variance as given before:

$$\hat{\beta} \sim \mathcal{N}\left(\beta, \sigma^2 (X^{\mathrm{T}} X)^{-1}\right)$$

where Q is the cofactor matrix. This estimator reaches the Cramér–Rao bound for the model, and thus is optimal in the class of all unbiased estimators. Note that unlike the Gauss–Markov theorem, this result establishes optimality among both linear and non-linear estimators, but only in the case of normally distributed error terms.

The estimator s^2 will be proportional to the chi-squared distribution:

$$s^2 \sim \frac{\sigma^2}{n-p} \cdot \chi^2_{n-p}$$

The variance of this estimator is equal to $2\sigma^4/(n-p)$, which does not attain the Cramér–Rao bound of $2\sigma^4/n$. However it was shown that there are no unbiased estimators of σ^2 with variance smaller than that of the estimator s^2. If we are willing to allow biased estimators, and consider the class of estimators that are proportional to the sum of squared residuals (SSR) of the model, then the best (in the sense of the mean squared error) estimator in this class will be $\tilde{\sigma}^2 = SSR/(n-p+2)$, which even beats the Cramér–Rao bound in case when there is only one regressor ($p = 1$).

Moreover, the estimators $\hat{\beta}$ and s^2 are independent, the fact which comes in useful when constructing the t- and F-tests for the regression.

Influential Observations

As was mentioned before, the estimator $\hat{\beta}$ is linear in y, meaning that it represents a linear combination of the dependent variables y_i. The weights in this linear combination are functions of the regressors X, and generally are unequal. The observations with high weights are called influential because they have a more pronounced effect on the value of the estimator.

To analyze which observations are influential we remove a specific j-th observation and consider how much the estimated quantities are going to change (similarly to the jackknife method). It can be shown that the change in the OLS estimator for β will be equal to,

$$\hat{\beta}^{(j)} - \hat{\beta} = -\frac{1}{1-h_j}(X^{\mathrm{T}}X)^{-1}x_j^{\mathrm{T}}\hat{\varepsilon}_j,$$

where $h_j = x_j^T \left(X^T X\right)^{-1} x_j$ is the j-th diagonal element of the hat matrix P, and x_j is the vector of regressors corresponding to the j-th observation. Similarly, the change in the predicted value for j-th observation resulting from omitting that observation from the dataset will be equal to,

$$\hat{y}_j^{(j)} - \hat{y}_j = x_j^{\mathrm{T}}\hat{\beta}^{(j)} - x_j^T\hat{\beta} = -\frac{h_j}{1-h_j}\hat{\varepsilon}_j$$

From the properties of the hat matrix, $0 \le h_j \le 1$, and they sum up to p, so that on average $h_j \approx p/n$. These quantities h_j are called the leverages, and observations with high h_j are called leverage points. Usually the observations with high leverage ought to be scrutinized more carefully, in case they are erroneous, or outliers, or in some other way atypical of the rest of the dataset.

Partitioned Regression

Sometimes the variables and corresponding parameters in the regression can be logically split into two groups, so that the regression takes form,

$$y = X_1\beta_1 + X_2\beta_2 + \varepsilon,$$

where X_1 and X_2 have dimensions $n\times p_1$, $n\times p_2$, and β_1, β_2 are $p_1\times 1$ and $p_2\times 1$ vectors, with $p_1 + p_2 = p$.

The Frisch–Waugh–Lovell theorem states that in this regression the residuals $\hat{\varepsilon}$ and the OLS estimate $\hat{\beta}_2$ will be numerically identical to the residuals and the OLS estimate for β_2 in the following regression:

$$M_1 y = M_1 X_2 \beta_2 + \eta,$$

where M_1 is the annihilator matrix for regressors X_1.

The theorem can be used to establish a number of theoretical results. For example, having a regression with a constant and another regressor is equivalent to subtracting the means from the dependent variable and the regressor and then running the regression for the de-meaned variables but without the constant term.

Constrained Estimation

Suppose it is known that the coefficients in the regression satisfy a system of linear equations,

$$A: \quad Q^T\beta = c,$$

where Q is a $p\times q$ matrix of full rank, and c is a $q\times 1$ vector of known constants, where $q < p$. In this case least squares estimation is equivalent to minimizing the sum of squared residuals of the model subject to the constraint A. The constrained least squares (CLS) estimator can be given by an explicit formula:

$$\hat{\beta}^c = \hat{\beta} - (X^TX)^{-1}Q\big(Q^T(X^TX)^{-1}Q\big)^{-1}(Q^T\hat{\beta} - c).$$

This expression for the constrained estimator is valid as long as the matrix X^TX is invertible. It was assumed from the beginning of this topic that this matrix is of full rank, and it was noted that when the rank condition fails, β will not be identifiable. However it may happen that adding the restriction A makes β identifiable, in which case one would like to find the formula for the estimator. The estimator is equal to,

$$\hat{\beta}^c = R(R^TX^TXR)^{-1}R^TX^Ty + \big(I_p - R(R^TX^TXR)^{-1}R^TX^TX\big)Q(Q^TQ)^{-1}c,$$

where R is a $p\times(p-q)$ matrix such that the matrix $[Q\ R]$ is non-singular, and $R^TQ = 0$. Such a matrix can always be found, although generally it is not unique. The second formula coincides with the first in case when X^TX is invertible.

Large Sample Properties

The least squares estimators are point estimates of the linear regression model parameters β. However, generally we also want to know how close those estimates might be to the true values of parameters. In other words, we want to construct the interval estimates.

Since we haven't made any assumption about the distribution of error term ε_i, it is impossible to infer the distribution of the estimators $\hat{\beta}$ and $\hat{\sigma}^2$. Nevertheless, we can apply the central limit theorem to derive their *asymptotic* properties as sample size n goes to infinity. While the sample size is necessarily finite, it is customary to assume that n is "large enough" so that the true distribution of the OLS estimator is close to its asymptotic limit.

We can show that under the model assumptions, the least squares estimator for β is consistent (that is $\hat{\beta}$ converges in probability to β) and asymptotically normal:

$$(\hat{\beta}-\beta)\xrightarrow{d}\mathcal{N}\left(0,\sigma^2 Q_{xx}^{-1}\right),$$

where $Q_{xx} = X^T X$

Intervals

Using this asymptotic distribution, approximate two-sided confidence intervals for the j-th component of the vector $\hat{\beta}$ can be constructed as:

$$\beta_j \in \left[\hat{\beta}_j \pm q_{1-\frac{\alpha}{2}}^{\mathcal{N}(0,1)}\sqrt{\hat{\sigma}^2\left[Q_{xx}^{-1}\right]_{jj}}\ \right] \text{ at the } 1-\alpha \text{ confidence level,}$$

where q denotes the quantile function of standard normal distribution, and $[\cdot]_{jj}$ is the j-th diagonal element of a matrix.

Similarly, the least squares estimator for σ^2 is also consistent and asymptotically normal (provided that the fourth moment of ε_i exists) with limiting distribution,

$$(\hat{\sigma}^2-\sigma^2)\xrightarrow{d}\mathcal{N}\left(0,\mathrm{E}\left[\varepsilon_i^4\right]-\sigma^4\right).$$

These asymptotic distributions can be used for prediction, testing hypotheses, constructing other estimators, etc. As an example consider the problem of prediction. Suppose x_0 is some point within the domain of distribution of the regressors, and one wants to know what the response variable would have been at that point. The mean response is the quantity $y_0 = x_0^T\beta$, whereas the predicted response is $\hat{y}_0 = x_0^T\hat{\beta}$. Clearly the predicted response is a random variable, its distribution can be derived from that of $\hat{\beta}$:

$$\left(\hat{y}_0 - y_0\right)\xrightarrow{d}\mathcal{N}\left(0,\sigma^2 x_0^T Q_{xx}^{-1} x_0\right),$$

which allows construct confidence intervals for mean response y_o to be constructed:

$$y_o \in \left[x_o^{T}\hat{\beta} \pm q_{1-\frac{\alpha}{2}}^{\mathcal{N}(0,1)} \sqrt{\hat{\sigma}^2 x_o^{T} Q_{xx}^{-1} x_o} \right] \text{ at the } 1-\alpha \text{ confidence level.}$$

Hypothesis Testing

Two hypothesis tests are particularly widely used. First, one wants to know if the estimated regression equation is any better than simply predicting that all values of the response variable equal its sample mean (if not, it is said to have no explanatory power). The null hypothesis of no explanatory value of the estimated regression is tested using an F-test. If the calculated F-value is found to be large enough to exceed its critical value for the pre-chosen level of significance, the null hypothesis is rejected and the alternative hypothesis, that the regression has explanatory power, is accepted. Otherwise, the null hypothesis of no explanatory power is accepted.

Second, for each explanatory variable of interest, one wants to know whether its estimated coefficient differs significantly from zero—that is, whether this particular explanatory variable in fact has explanatory power in predicting the response variable. Here the null hypothesis is that the true coefficient is zero. This hypothesis is tested by computing the coefficient's t-statistic, as the ratio of the coefficient estimate to its standard error. If the t-statistic is larger than a predetermined value, the null hypothesis is rejected and the variable is found to have explanatory power, with its coefficient significantly different from zero. Otherwise, the null hypothesis of a zero value of the true coefficient is accepted.

In addition, the Chow test is used to test whether two subsamples both have the same underlying true coefficient values. The sum of squared residuals of regressions on each of the subsets and on the combined data set are compared by computing an F-statistic; if this exceeds a critical value, the null hypothesis of no difference between the two subsets is rejected; otherwise, it is accepted.

Example with Real Data

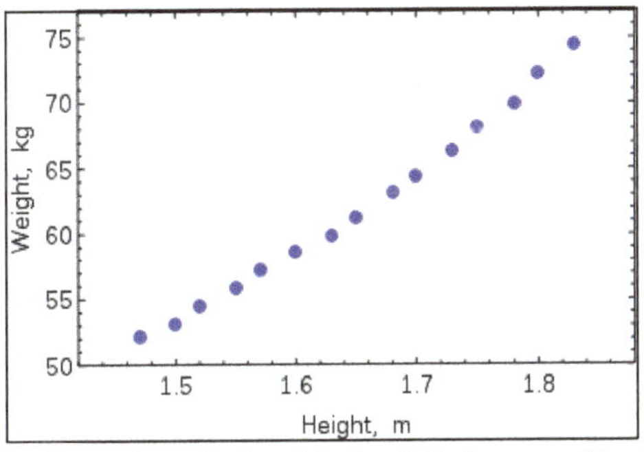

Scatterplot of the data, the relationship is slightly curved but close to linear.

The following data set gives average heights and weights for American women aged 30–39.

Height (m)	1.47	1.50	1.52	1.55	1.57	1.60	1.63	1.65	1.68	1.70	1.73	1.75	1.78	1.80	1.83
Weight (kg)	52.21	53.12	54.48	55.84	57.20	58.57	59.93	61.29	63.11	64.47	66.28	68.10	69.92	72.19	74.46

When only one dependent variable is being modeled, a scatterplot will suggest the form and strength of the relationship between the dependent variable and regressors. It might also reveal outliers, heteroscedasticity, and other aspects of the data that may complicate the interpretation of a fitted regression model. The scatterplot suggests that the relationship is strong and can be approximated as a quadratic function. OLS can handle non-linear relationships by introducing the regressor height. The regression model then becomes a multiple linear model:

$$w_i = \beta_1 + \beta_2 h_i + \beta_3 h_i^2 + \varepsilon_i.$$

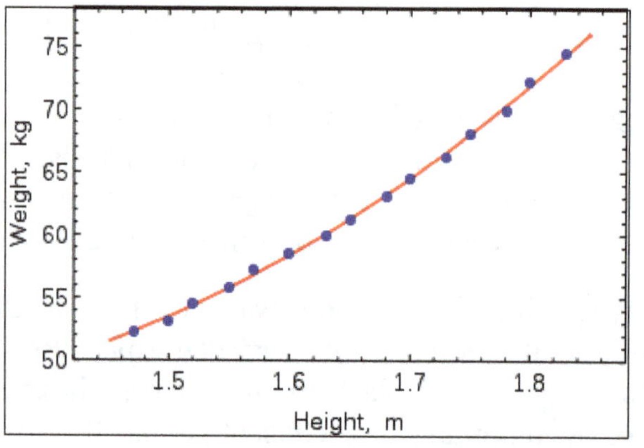

Fitted Regression

The output from most popular statistical packages will look similar to this:

Method	Least squares			
Dependent variable	WEIGHT			
Observations	15			
Parameter	**Value**	**Std error**	**t-statistic**	**p-value**
β_1	128.8128	16.3083	7.8986	0.0000
β_2	−143.1620	19.8332	−7.2183	0.0000
β_3	61.9603	6.0084	10.3122	0.0000
R^2	0.9989	S.E. of regression		0.2516
Adjusted R^2	0.9987	Model sum-of-sq.		692.61
Log-likelihood	1.0890	Residual sum-of-sq.		0.7595
Durbin–Watson stat.	2.1013	Total sum-of-sq.		693.37
Akaike criterion	0.2548	F-statistic		5471.2
Schwarz criterion	0.3964	p-value (F-stat)		0.0000

In this table:

- The *Value* column gives the least squares estimates of parameters β_j.
- The *Std error* column shows standard errors of each coefficient estimate: $\hat{\sigma}_j = \left(\hat{\sigma}^2 \left[Q_{xx}^{-1}\right]_{jj}\right)^{\frac{1}{2}}$.
- The *t-statistic* and *p-value* columns are testing whether any of the coefficients might be equal to zero. The *t*-statistic is calculated simply as $t = \hat{\beta}_j / \hat{\sigma}_j$. If the errors ε follow a normal distribution, *t* follows a Student-t distribution. Under weaker conditions, *t* is

asymptotically normal. Large values of *t* indicate that the null hypothesis can be rejected and that the corresponding coefficient is not zero. The second column, *p*-value, expresses the results of the hypothesis test as a significance level. Conventionally, *p*-values smaller than 0.05 are taken as evidence that the population coefficient is nonzero.

- *R-squared* is the coefficient of determination indicating goodness-of-fit of the regression. This statistic will be equal to one if fit is perfect, and to zero when regressors X have no explanatory power whatsoever. This is a biased estimate of the population *R-squared*, and will never decrease if additional regressors are added, even if they are irrelevant.
- *Adjusted R-squared* is a slightly modified version of R^2, designed to penalize for the excess number of regressors which do not add to the explanatory power of the regression. This statistic is always smaller than R^2, can decrease as new regressors are added, and even be negative for poorly fitting models:

$$\bar{R}^2 = 1 - \frac{n-1}{n-p}(1-R^2)$$

- *Log-likelihood* is calculated under the assumption that errors follow normal distribution. Even though the assumption is not very reasonable, this statistic may still find its use in conducting LR tests.
- *Durbin–Watson statistic* tests whether there is any evidence of serial correlation between the residuals. As a rule of thumb, the value smaller than 2 will be an evidence of positive correlation.
- *Akaike information criterion* and *Schwarz criterion* are both used for model selection. Generally when comparing two alternative models, smaller values of one of these criteria will indicate a better model.
- *Standard error of regression* is an estimate of σ, standard error of the error term.
- *Total sum of squares, model sum of squared,* and *residual sum of squares* tell us how much of the initial variation in the sample were explained by the regression.
- *F-statistic* tries to test the hypothesis that all coefficients (except the intercept) are equal to zero. This statistic has $F(p-1, n-p)$ distribution under the null hypothesis and normality assumption, and its *p-value* indicates probability that the hypothesis is indeed true. Note that when errors are not normal this statistic becomes invalid, and other tests such as Wald test or LR test should be used.

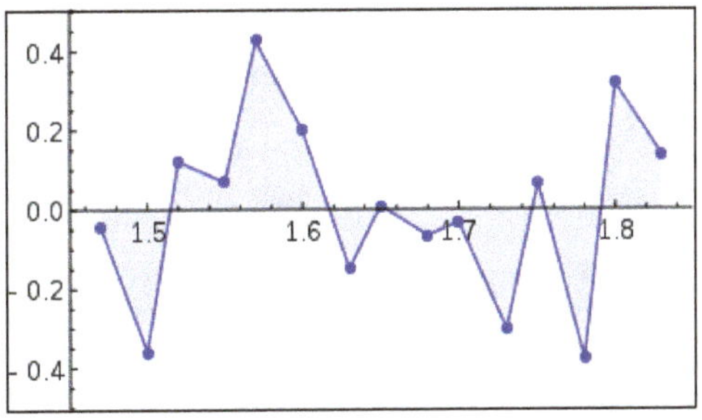

Residuals Plot

Ordinary least squares analysis often includes the use of diagnostic plots designed to detect departures of the data from the assumed form of the model. These are some of the common diagnostic plots:

- Residuals against the explanatory variables in the model. A non-linear relation between these variables suggests that the linearity of the conditional mean function may not hold. Different levels of variability in the residuals for different levels of the explanatory variables suggests possible heteroscedasticity.
- Residuals against explanatory variables not in the model. Any relation of the residuals to these variables would suggest considering these variables for inclusion in the model.
- Residuals against the fitted values, $\hat{y}$.
- Residuals against the preceding residual. This plot may identify serial correlations in the residuals.

An important consideration when carrying out statistical inference using regression models is how the data were sampled. In this example, the data are averages rather than measurements on individual women. The fit of the model is very good, but this does not imply that the weight of an individual woman can be predicted with high accuracy based only on her height.

Sensitivity to Rounding

This example also demonstrates that coefficients determined by these calculations are sensitive to how the data is prepared. The heights were originally given rounded to the nearest inch and have been converted and rounded to the nearest centimeter. Since the conversion factor is one inch to 2.54 cm this is *not* an exact conversion. The original inches can be recovered by Round(x/0.0254) and then re-converted to metric without rounding. If this is done the results become:

	Const	Height	Height²
Converted to metric with rounding.	128.8128	−143.162	61.96033
Converted to metric without rounding.	119.0205	−131.5076	58.5046

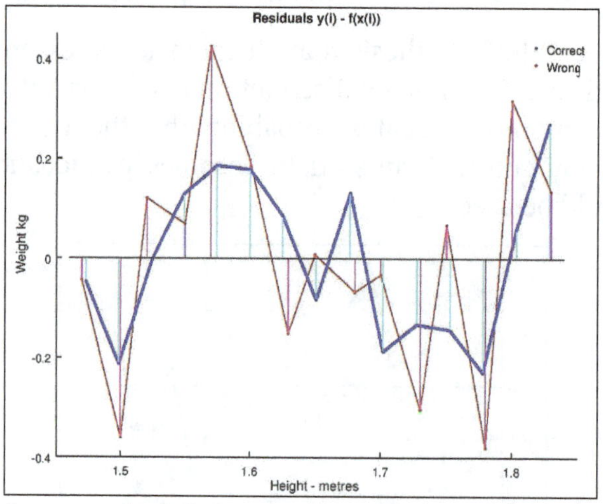

Residuals to a quadratic fit for correctly and incorrectly converted data.

Using either of these equations to predict the weight of a 5' 6" (1.6764m) woman gives similar values: 62.94 kg with rounding vs. 62.98 kg without rounding. Thus a seemingly small variation in the data has a real effect on the coefficients but a small effect on the results of the equation.

While this may look innocuous in the middle of the data range it could become significant at the extremes or in the case where the fitted model is used to project outside the data range (extrapolation).

This highlights a common error: this example is an abuse of OLS which inherently requires that the errors in the independent variable (in this case height) are zero or at least negligible. The initial rounding to nearest inch plus any actual measurement errors constitute a finite and non-negligible error. As a result, the fitted parameters are not the best estimates they are presumed to be. Though not totally spurious the error in the estimation will depend upon relative size of the x and y errors.

Another Example with Less Real Data

Problem Statement

We can use the least square mechanism to figure out the equation of a two body orbit in polar base co-ordinates. The equation typically used is $r(\theta)=\frac{p}{1-e\cos(\theta)}$ where $r(\theta)$ is the radius of how far the object is from one of the bodies. In the equation the parameters p and e are used to determine the path of the orbit. We have measured the following data.

θ (in degrees)	43	45	52	93	108	116
$r(\theta)$	4.7126	4.5542	4.0419	2.2187	1.8910	1.7599

We need to find the least-squares approximation of e and p for the given data.

Solution:

First we need to represent e and p in a linear form. So we are going to rewrite the equation $r(\theta)$ as $\frac{1}{r(\theta)}=\frac{1}{p}-\frac{e}{p}\cos(\theta)$. Now we can use this form to represent our observational data as:

$A^T A\begin{pmatrix} x \\ y \end{pmatrix}=A^T b$ where x is $\frac{1}{p}$ and y is $\frac{e}{p}$ and A is constructed by the first column being the coefficient of $\frac{1}{p}$ and the second column being the coefficient of $\frac{e}{p}$ and b is the values for the respective $\frac{1}{r(\theta)}$ so

$$A=\begin{bmatrix} 1 & -0.731354 \\ 1 & -0.707107 \\ 1 & -0.615661 \\ 1 & 0.052336 \\ 1 & 0.309017 \\ 1 & 0.438371 \end{bmatrix} \text{ and } b=\begin{bmatrix} 0.21220 \\ 0.21958 \\ 0.24741 \\ 0.45071 \\ 0.52883 \\ 0.56820 \end{bmatrix}.$$

On solving we get $\begin{pmatrix} x \\ y \end{pmatrix} = \begin{pmatrix} 0.43478 \\ 0.30435 \end{pmatrix}$

so $p = \frac{1}{x} = 2.3000 \text{ and } e = p \cdot y = 0.70001$

Probit Model

Probit regression, also called a probit model, is used to model dichotomous or binary outcome variables. In the probit model, the inverse standard normal distribution of the probability is modeled as a linear combination of the predictors.

```
+/- R Code

library(aod)

library(ggplot2)
```

Examples

- Example 1. Suppose that we are interested in the factors that influence whether a political candidate wins an election. The outcome (response) variable is binary (0/1); win or lose. The predictor variables of interest are the amount of money spent on the campaign, the amount of time spent campaigning negatively and whether or not the candidate is an incumbent.
- Example 2. A researcher is interested in how variables, such as GRE (Graduate Record Exam scores), GPA (grade point average) and prestige of the undergraduate institution, effect admission into graduate school. The response variable, admit/don't admit, is a binary variable.

Theoretical Background

Probit regression is a special type of the Generalized Linear Models (GLM; will be explained later). Here, the bivariate outcome Y has a Bernoulli distribution with parameter p (success probability $p \in (0,1)$). Recall that EY=p The *probit link function*,

$$probit(EY) = \Phi^{-1}(p) = \Phi^{-1}(p[Y=1])$$

is used to transform the expectation of this 0/1 dependent variable. Then, the probit of the mean is modelled as a linear combination of the covariates (regressors) X, i.e., we have a *linear predictor*

$$probit(EY) = \mathbf{X}\beta,$$

where β is a vector of unknown parameters. The *maximum likelihood* based approach is used for the parameter estimation, where a version of the *IRLS algorithm* is applied .

Predicted probability can be obtained by the *inverse probit* (i.e., standard normal cdf) transformation

$$\hat{P}[Y_i = 1] = \Phi(\mathbf{X}_{i\cdot}\hat{\beta}).$$

References

- Gelman, Andrew; Hill, Jennifer (2007). Data Analysis Using Regression and Multilevel/Hierarchical Models. New York: Cambridge University Press. pp. 79–108. ISBN 978-0-521-68689-1
- What-is-linear-regression: statisticssolutions.com, Retrieved 26, February 2020
- Grundel, R.; Pavlovic, N. B. (2007). "Response of bird species densities to habitat structure and fire history along a Midwestern open–forest Gradient". The Condor. 109 (4): 734–749. doi:10.1650/0010-5422(2007)109 [734:ROBSDT]2.0.CO;2
- Semiparametric: statisticshowto.datasciencecentral.com, Retrieved 25, April 2020
- Ver Hoef, JAY M.; Boveng, Peter L. (2007-01-01). "Quasi-Poisson vs. Negative Binomial Regression: How should we model overdispersed count data?". Ecology. 88 (11): 2766–2772. doi:10.1890/07-0043.1. Retrieved 17, January 2020
- Regression-analysis: businessjargons.com, Retrieved 14, March 2020
- DeBano, S. J.; Hamm, P. B.; Jensen, A.; Rondon, S. I.; Landolt, P. J. (2010). "Spatial and temporal dynamics of potato tuberworm (Lepidoptera: Gelechiidae) in the Columbia Basin of the Pacific Northwest". Environmental Entomology. 39 (1): 1–14. doi:10.1603/EN08270
- Multinomial-logistic-regression: statisticshowto.datasciencecentral.com, Retrieved 21, August 2020
- Harrell, Frank E. (2015). Regression Modeling Strategies. Springer Series in Statistics (2nd ed.). New York; Springer. doi:10.1007/978-3-319-19425-7. ISBN 978-3-319-19424-0
- Comprehensive-guide-regression: analyticsvidhya.com, Retrieved 06, June 2020

7

Stochastic Processes

The collection of random variables that are guided by a mathematical set which means every random variable is incomparably associated with an element in the set is known as stochastic process. This chapter includes Wiener process, Ornstein-Uhlenbeck process, Random walk and Poisson point process. It will help the reader in gaining a better perspective about the stochastic process.

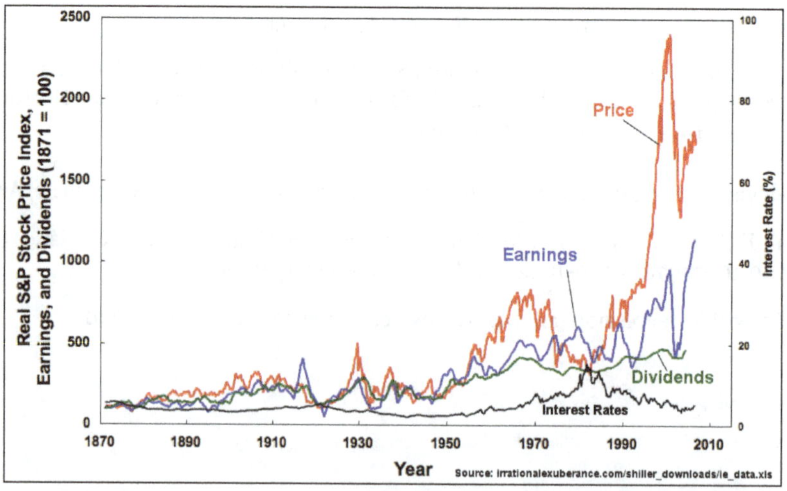

Stock market fluctuations have been modeled by stochastic processes.

A stochastic process (or random process) is a probability model used to describe phenomena that evolve over time or space. More specifically, in probability theory, a stochastic process is a time sequence representing the evolution of some system represented by a variable whose change is subject to a random variation. This is the probabilistic counterpart to a deterministic process (or deterministic system). Instead of describing a process which can only evolve in one way (as in the case, for example, of solutions of an ordinary differential equation), in a stochastic, or random process, there is some indeterminacy: even if the initial condition (or starting point) is known, there are several (often infinitely many) directions in which the process may evolve. In many stochastic processes, the movement to the next state or position depends on only the current state, and is independent from prior states or values the process has taken.

In the simple case of discrete time, as opposed to continuous time, a stochastic process is a sequence of random variables. The random variables corresponding to various times may be completely different, the only requirement being that these different random quantities all take values in the same space (the codomain of the function). One approach may be to model these random variables as random functions of one or several deterministic arguments (in most cases, the time parameter). Although the random values of a stochastic process at different times may be independent random variables, in most commonly considered situations they exhibit complicated statistical dependence.

Familiar examples of stochastic processes include stock market and exchange rate fluctuations; signals such as speech; audio and video; medical data such as a patient's EKG, EEG, blood pressure or temperature; and random movement such as Brownian motion or random walks.

A generalization, the random field, is defined by letting the variables be parametrized by members of a topological space instead of time. Examples of random fields include static images, random terrain (landscapes), wind waves and composition variations of a heterogeneous material.

Formal Definition and Basic Properties

Definition

Given a probability space $(\Omega, \mathcal{F}, P)$ and a measurable space (S, Σ), an S-valued stochastic process is a collection of S-valued random variables on Ω, indexed by a totally ordered set T ("time"). That is, a stochastic process X is a collection,

$$\{X_t : t \in T\}$$

where each X_t is an S-valued random variable on Ω. The space S is then called the state space of the process.

Finite-dimensional Distributions

Let X be an S-valued stochastic process. For every finite sequence $T' = (t_1, \ldots, t_k) \in T^k$, the k-tuple $X_{T'} = (X_{t_1}, X_{t_2}, \ldots, X_{t_k})$ is a random variable taking values in S^k. The distribution $\mathbb{P}_{T'}(\cdot) = \mathbb{P}(X_{T'}^{-1}(\cdot))$ of this random variable is a probability measure on S^k. This is called a finite-dimensional distribution of X.

Under suitable topological restrictions, a suitably "consistent" collection of finite-dimensional distributions can be used to define a stochastic process.

History of Stochastic Processes

Stochastic processes were first studied rigorously in the late 19th century to aid in understanding financial markets and Brownian motion. The first person to describe the mathematics behind Brownian motion was Thorvald N. Thiele in a paper on the method of least squares published in 1880. This was followed independently by Louis Bachelier in 1900 in his PhD thesis "The theory of speculation", in which he presented a stochastic analysis of the stock and option markets. Albert Einstein (in one of his 1905 papers) and Marian Smoluchowski (1906) brought the solution of the problem to the attention of physicists, and presented it as a way to indirectly confirm the existence of atoms and molecules. Their equations describing Brownian motion were subsequently verified by the experimental work of Jean Baptiste Perrin in 1908.

An excerpt from Einstein's paper describes the fundamentals of a stochastic model:

"It must clearly be assumed that each individual particle executes a motion which is independent of the motions of all other particles; it will also be considered that the movements of one and the same particle in different time intervals are independent processes, as long as these time intervals are not chosen too small.

We introduce a time interval τ into consideration, which is very small compared to the observable time intervals, but nevertheless so large that in two successive time intervals τ, the motions executed by the particle can be thought of as events which are independent of each other".

Construction

In the ordinary axiomatization of probability theory by means of measure theory, the problem is to construct a sigma-algebra of measurable subsets of the space of all functions, and then put a finite measure on it. For this purpose one traditionally uses a method called Kolmogorov extension.

Kolmogorov Extension

The Kolmogorov extension proceeds along the following lines: assuming that a probability measure on the space of all functions $f: X \to Y$ exists, then it can be used to specify the joint probability distribution of finite-dimensional random variables $f(x_1),\ldots,f(x_n)$. Now, from this n-dimensional probability distribution we can deduce an (n – 1)-dimensional marginal probability distribution for $f(x_1),\ldots,f(x_{n-1})$. Note that the obvious compatibility condition, namely, that this marginal probability distribution be in the same class as the one derived from the full-blown stochastic process, is not a requirement. Such a condition only holds, for example, if the stochastic process is a Wiener process (in which case the marginals are all gaussian distributions of the exponential class) but not in general for all stochastic processes. When this condition is expressed in terms of probability densities, the result is called the Chapman–Kolmogorov equation.

The Kolmogorov extension theorem guarantees the existence of a stochastic process with a given family of finite-dimensional probability distributions satisfying the Chapman–Kolmogorov compatibility condition.

Separability, or What the Kolmogorov Extension Does Not Provide

Recall that in the Kolmogorov axiomatization, measurable sets are the sets which have a probability or, in other words, the sets corresponding to yes/no questions that have a probabilistic answer.

The Kolmogorov extension starts by declaring to be measurable all sets of functions where finitely many coordinates $[f(x_1),\ldots,f(x_n)]$ are restricted to lie in measurable subsets of Y_n. In other words, if a yes/no question about f can be answered by looking at the values of at most finitely many coordinates, then it has a probabilistic answer.

In measure theory, if we have a countably infinite collection of measurable sets, then the union and intersection of all of them is a measurable set. For our purposes, this means that yes/no questions that depend on countably many coordinates have a probabilistic answer.

The good news is that the Kolmogorov extension makes it possible to construct stochastic processes with fairly arbitrary finite-dimensional distributions. Also, every question that one could ask about a sequence has a probabilistic answer when asked of a random sequence. The bad news is that certain questions about functions on a continuous domain don't have a probabilistic answer. One might hope that the questions that depend on uncountably many values of a function be of little interest, but the really bad news is that virtually all concepts of calculus are of this sort. For example:

1. boundedness
2. continuity
3. differentiability

all require knowledge of uncountably many values of the function.

One solution to this problem is to require that the stochastic process be separable. In other words, that there be some countable set of coordinates $\{f(x_i)\}$ whose values determine the whole random function *f*.

The Kolmogorov continuity theorem guarantees that processes that satisfy certain constraints on the moments of their increments have continuous modifications and are therefore separable.

Filtrations

Given a probability space $(\Omega, \mathcal{F}, P)$, a filtration is a weakly increasing collection of sigma-algebras on Ω, $\{\mathcal{F}_t, t \in T\}$, indexed by some totally ordered set T, and bounded above by $\mathcal{F}$, i.e. for *s,t* $\in T$ with *s* < *t*,

$$\mathcal{F}_s \subseteq \mathcal{F}_t \subseteq \mathcal{F}.$$

A stochastic process X on the same time set T is said to be adapted to the filtration if, for every t $\in T$, X_t is $\mathcal{F}_t$-measurable.

Natural Filtration

Given a stochastic process $X = \{X_t : t \in T\}$,, the natural filtration for (or induced by) this process is the filtration where $\mathcal{F}_t$ is generated by all values of X_s up to time *s* = *t*, i.e. $\mathcal{F}_t = \sigma(\{X_s^{-1}(A) : s \leq t, A \in \Sigma\})$.

A stochastic process is always adapted to its natural filtration.

Classification

Stochastic processes can be classified according to the cardinality of its index set (usually interpreted as time) and state space.

Discrete Time and Discrete State Space

If both t and X_t belong to $\mathbb{N}$, the set of natural numbers, then we have models which lead to Markov chains. For example:

(a) If X_t means the bit (0 or 1) in position t of a sequence of transmitted bits, then X_t can be modelled as a Markov chain with two states. This leads to the error-correcting Viterbi algorithm in data transmission.

(b) If X_t represents the combined genotype of a breeding couple in the tth generation in an inbreeding model, it can be shown that the proportion of heterozygous individuals in the population approaches zero as t goes to ∞.

Continuous Time and Continuous State Space

The paradigm of continuous stochastic process is that of the Wiener process. In its original form the problem was concerned with a particle floating on a liquid surface, receiving "kicks" from the molecules of the liquid. The particle is then viewed as being subject to a random force which, since the molecules are very small and very close together, is treated as being continuous and since the particle is constrained to the surface of the liquid by surface tension, is at each point in time a vector parallel to the surface. Thus, the random force is described by a two-component stochastic process; two real-valued random variables are associated to each point in the index set, time, (note that since the liquid is viewed as being homogeneous the force is independent of the spatial coordinates) with the domain of the two random variables being R, giving the x and y components of the force. A treatment of Brownian motion generally also includes the effect of viscosity, resulting in an equation of motion known as the Langevin equation.

Discrete Time and Continuous State Space

If the index set of the process is N (the natural numbers), and the range is R (the real numbers), there are some natural questions to ask about the sample sequences of a process $\{X_i\}_{i \in N}$, where a sample sequence is $\{X_i(\omega)\}_{i \in N}$.

1. What is the probability that each sample sequence is bounded?
2. What is the probability that each sample sequence is monotonic?
3. What is the probability that each sample sequence has a limit as the index approaches ∞?
4. What is the probability that the series obtained from a sample sequence from converges?
5. What is the probability distribution of the sum?

Main applications of discrete time continuous state stochastic models include Markov chain Monte Carlo (MCMC) and the analysis of Time Series.

Continuous Time and Discrete State Space

Similarly, if the index space I is a finite or infinite interval, we can ask about the sample paths $\{X_t(\omega)\}_{t \in I}$

1. What is the probability that it is bounded/integrable...?
2. What is the probability that it has a limit at ∞
3. What is the probability distribution of the integral?

Wiener Process

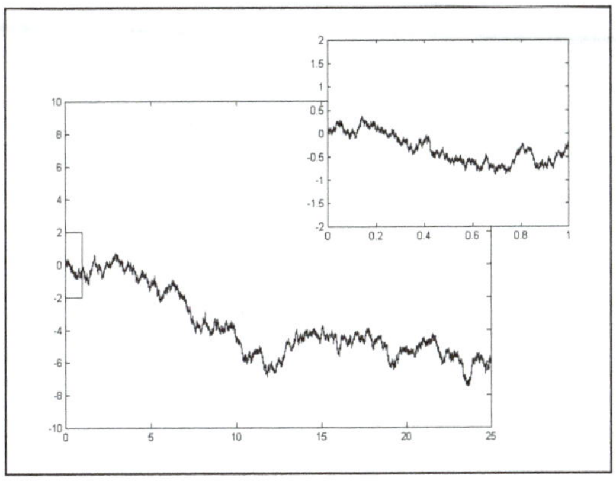

A single realization of a one-dimensional Wiener process.

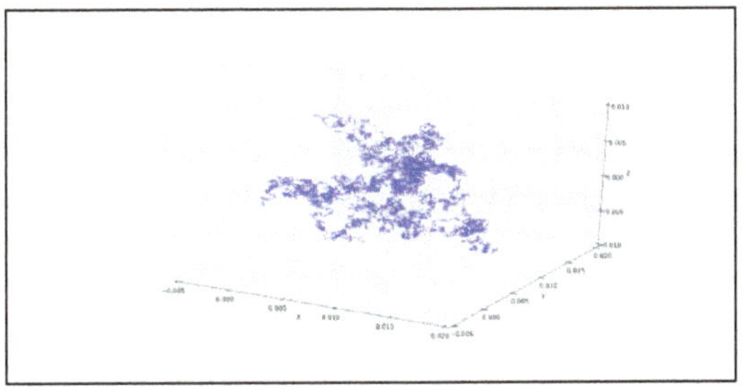

A single realization of a three-dimensional Wiener process.

In mathematics, the Wiener process is a continuous-time stochastic process named in honor of Norbert Wiener. It is often called standard Brownian motion, after Robert Brown. It is one of the best known Lévy processes (càdlàg stochastic processes with stationary independent increments) and occurs frequently in pure and applied mathematics, economics, quantitative finance, and physics.

The Wiener process plays an important role both in pure and applied mathematics. In pure mathematics, the Wiener process gave rise to the study of continuous time martingales. It is a key process in terms of which more complicated stochastic processes can be described. As such, it plays a vital role in stochastic calculus, diffusion processes and even potential theory. It is the driving process of Schramm–Loewner evolution. In applied mathematics, the Wiener process is used to represent the integral of a white noise Gaussian process, and so is useful as a model of noise in electronics engineering, instrument errors in filtering theory and unknown forces in control theory.

The Wiener process has applications throughout the mathematical sciences. In physics it is used to study Brownian motion, the diffusion of minute particles suspended in fluid, and other types of diffusion via the Fokker–Planck and Langevin equations. It also forms the basis for the rigorous path integral formulation of quantum mechanics (by the Feynman–Kac formula, a solution to the Schrödinger equation can be represented in terms of the Wiener process) and the study of eternal inflation in physical cosmology. It is also prominent in the mathematical theory of finance, in particular the Black–Scholes option pricing model.

Characterisations of the Wiener process

The Wiener process W_t is characterised by the following properties:

1. $W_0 = 0$ a.s.
2. W has independent increments: $W_{t+u} - W_t$ is independent of $\sigma(W_s : s \le t)$ for $u \ge 0$
3. W has Gaussian increments: $W_{t+u} - W_t$ is normally distributed with mean 0 and variance u, $W_{t+u} - W_t \sim N(0, u)$
4. W has continuous paths: With probability 1, W_t is continuous in t.

The independent increments means that if $0 \le s_1 < t_1 \le s_2 < t_2$ then $W_{t1} - W_{s1}$ and $W_{t2} - W_{s2}$ are independent random variables, and the similar condition holds for n increments.

An alternative characterisation of the Wiener process is the so-called *Lévy characterisation* that says that the Wiener process is an almost surely continuous martingale with $W_0 = 0$ and quadratic variation $[W_t, W_t] = t$ (which means that $W_t^2 - t$ is also a martingale).

A third characterisation is that the Wiener process has a spectral representation as a sine series whose coefficients are independent $N(0, 1)$ random variables. This representation can be obtained using the Karhunen–Loève theorem.

Another characterisation of a Wiener process is the Definite integral (from zero to time t) of a zero mean, unit variance, delta correlated ("white") Gaussian process.

The Wiener process can be constructed as the scaling limit of a random walk, or other discrete-time stochastic processes with stationary independent increments. This is known as Donsker's theorem. Like the random walk, the Wiener process is recurrent in one or two dimensions (meaning that it returns almost surely to any fixed neighborhood of the origin infinitely often) whereas it is not recurrent in dimensions three and higher. Unlike the random walk, it is scale invariant, meaning that,

$$\alpha^{-1} W_{\alpha^2 t}$$

is a Wiener process for any nonzero constant α. The Wiener measure is the probability law on the space of continuous functions g, with $g(0) = 0$, induced by the Wiener process. An integral based on Wiener measure may be called a Wiener integral.

Wiener Process as a Limit of Random Walk

Let $\xi_1, \xi_2, \ldots$ be i.i.d. random variables with mean 0 and variance 1. For each n, define a continuous time stochastic process,

$$W_n(t) = \frac{1}{\sqrt{n}} \sum_{1 \le k \le \lfloor nt \rfloor} \xi_k$$

This is a random step function. Increments of W_n are independent because the are independent. For large n, $W_n(t) - W_n(s)$ is close to $N(0, t-s)$ by the central limit theorem. It's tempting to believe that as $n \to \infty$, W_n will approach a Wiener process, and this temptation bears true fruit. The proof is provided by Donsker's theorem. This formulation explained why Brownian motion is ubiquitous.

Properties of a One-dimensional Wiener Process

Basic Properties

The unconditional probability density function, which follows normal distribution with mean = 0 and variance = *t*, at a fixed time *t*:

$$f_{W_t}(x) = \frac{1}{\sqrt{2\pi t}} e^{-x^2/(2t)}.$$

The expectation is zero:

$$E[W_t] = 0.$$

The variance, using the computational formula, is *t*:

$$\text{Var}(W_t) = E\left[W_t^2\right] - E^2[W_t] = E\left[W_t^2\right] - 0 = E\left[W_t^2\right] = t.$$

Covariance and Correlation

The covariance and correlation:

$$\text{cov}(W_s, W_t) = \min(s,t),$$

$$\text{corr}(W_s, W_t) = \frac{\text{cov}(W_s, W_t)}{\sigma_{W_s}\sigma_{W_t}} = \frac{\min(s,t)}{\sqrt{st}} = \sqrt{\frac{\min(s,t)}{\max(s,t)}}.$$

The results for the expectation and variance follow immediately from the definition that increments have a normal distribution, centered at zero. Thus:

$$W_t = W_t - W_0 \sim N(0,t).$$

The results for the covariance and correlation follow from the definition that non-overlapping increments are independent, of which only the property that they are uncorrelated is used. Suppose that $t_1 < t_2$.

$$\text{cov}(W_{t_1}, W_{t_2}) = \text{E}\left[(W_{t_1} - \text{E}[W_{t_1}]) \cdot (W_{t_2} - \text{E}[W_{t_2}])\right] = \text{E}\left[W_{t_1} \cdot W_{t_2}\right].$$

Substituting

$$W_{t_2} = (W_{t_2} - W_{t_1}) + W_{t_1}$$

we arrive at:

$$\begin{aligned} \text{E}[W_{t_1} \cdot W_{t_2}] &= \text{E}\left[W_{t_1} \cdot ((W_{t_2} - W_{t_1}) + W_{t_1})\right] \\ &= \text{E}\left[W_{t_1} \cdot (W_{t_2} - W_{t_1})\right] + \text{E}\left[W_{t_1}^2\right]. \end{aligned}$$

Since $W(t_1) = W(t_1) - W(t_0)$ and $W(t_2) - W(t_1)$, are independent,

$$\mathrm{E}\left[W_{t_1}\cdot(W_{t_2}-W_{t_1})\right]=\mathrm{E}[W_{t_1}]\cdot\mathrm{E}[W_{t_2}-W_{t_1}]=0.$$

Thus:

$$\operatorname{cov}(W_{t_1},W_{t_2})=\mathrm{E}\left[W_{t_1}^2\right]=t_1.$$

Wiener Representation

Wiener (1923) also gave a representation of a Brownian path in terms of a random Fourier series. If ξ_n are independent Gaussian variables with mean zero and variance one, then:

$$W_t=\xi_0 t+\sqrt{2}\sum_{n=1}^{\infty}\xi_n\frac{\sin\pi nt}{\pi n}$$

and

$$W_t=\sqrt{2}\sum_{n=1}^{\infty}\xi_n\frac{\sin\left(\left(n-\frac{1}{2}\right)\pi t\right)}{\left(n-\frac{1}{2}\right)\pi}$$

represent a Brownian motion on $[0,1]$. The scaled process:

$$\sqrt{c}W\left(\frac{t}{c}\right)$$

is a Brownian motion on $[0,c]$ (cf. Karhunen–Loève theorem).

Running Maximum

The joint distribution of the running maximum:

$$M_t=\max_{0\le s\le t}W_s$$

and W_t is

$$f_{M_t,W_t}(m,w)=\frac{2(2m-w)}{t\sqrt{2\pi t}}e^{-\frac{(2m-w)^2}{2t}},\qquad m\ge 0, w\le m.$$

To get the unconditional distribution of f_{M_t}, integrate over $-\infty < w \le m$:

$$f_{M_t}(m)=\int_{-\infty}^{m}f_{M_t,W_t}(m,w)dw=\int_{-\infty}^{m}\frac{2(2m-w)}{t\sqrt{2\pi t}}e^{-\frac{(2m-w)^2}{2t}}dw$$

$$=\sqrt{\frac{2}{\pi t}}e^{-\frac{m^2}{2t}},\qquad m\ge 0.$$

And the expectation

$$\mathrm{E}[M_t] = \int_0^\infty m f_{M_t}(m)\,dm = \int_0^\infty m\sqrt{\frac{2}{\pi t}}e^{-\frac{m^2}{2t}}\,dm = \sqrt{\frac{2t}{\pi}}$$

Self-similarity

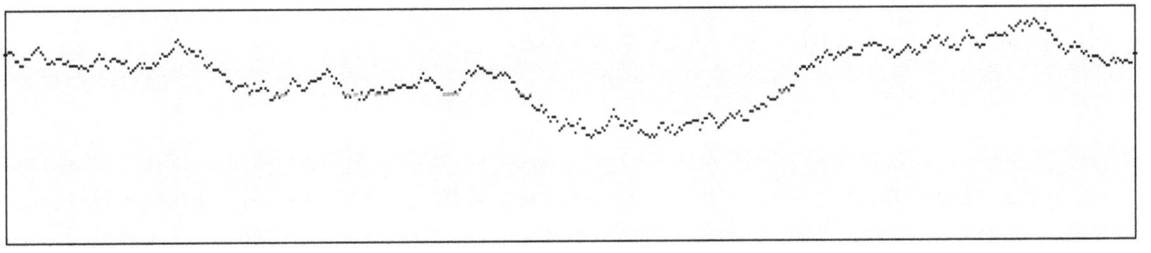

A demonstration of Brownian scaling, showing $V_t = (1/\sqrt{c})W_{ct}$ for decreasing c. Note that the average features of the function do not change while zooming in, and note that it zooms in quadratically faster horizontally than vertically.

Brownian Scaling

For every $c > 0$ the process $V_t = (1/\sqrt{c})W_{ct}$ is another Wiener process.

Time Reversal

The process $V_t = W_1 - W_{1-t}$ for $0 \leq t \leq 1$ is distributed like W_t for $0 \leq t \leq 1$.

Time Inversion

The process $V_t = tW_{1/t}$ is another Wiener process.

A class of Brownian Martingales

If a polynomial $p(x, t)$ satisfies the PDE

$$\left(\frac{\partial}{\partial t} + \frac{1}{2}\frac{\partial^2}{\partial x^2}\right)p(x,t) = 0$$

then the stochastic process:

$$M_t = p(W_t, t)$$

is a martingale.

Example: $W_t^2 - t$ is a martingale, which shows that the quadratic variation of W on $[0, t]$ is equal to t. It follows that the expected time of first exit of W from $(-c, c)$ is equal to c^2.

More generally, for every polynomial $p(x, t)$ the following stochastic process is a martingale:

$$M_t = p(W_t, t) - \int_0^t a(W_s, s)ds,$$

where a is the polynomial:

$$a(x,t) = \left(\frac{\partial}{\partial t} + \frac{1}{2}\frac{\partial^2}{\partial x^2}\right)p(x,t).$$

Example: $p(x,t) = (x^2 - t)^2$, $a(x,t) = 4x^2$; the process,

$$(W_t^2 - t)^2 - 4\int_0^t W_s^2\, ds$$

is a martingale, which shows that the quadratic variation of the martingale $W_t^2 - t$ on [0, t] is equal to

$$4\int_0^t W_s^2\, ds.$$

About functions $p(xa, t)$ more general than polynomials.

Some Properties of Sample Paths

The set of all functions w with these properties is of full Wiener measure. That is, a path (sample function) of the Wiener process has all these properties almost surely.

Qualitative Properties

- For every ε > 0, the function w takes both (strictly) positive and (strictly) negative values on (0, ε).
- The function w is continuous everywhere but differentiable nowhere (like the Weierstrass function).
- Points of local maximum of the function w are a dense countable set; the maximum values are pairwise different; each local maximum is sharp in the following sense: if w has a local maximum at t then:

  $$\lim_{s \to t} \frac{|w(s) - w(t)|}{|s - t|} \to \infty.$$

  The same holds for local minima.
- The function w has no points of local increase, that is, no t > 0 satisfies the following for some ε in (0, t): first, $w(s) \le w(t)$ for all s in (t – ε, t), and second, $w(s) \ge w(t)$ for all s in (t, t + ε). (Local increase is a weaker condition than that w is increasing on (t – ε, t + ε).) The same holds for local decrease.
- The function w is of unbounded variation on every interval.

- The quadratic variation of w over [0,t] is t.
- Zeros of the function w are a nowhere dense perfect set of Lebesgue measure 0 and Hausdorff dimension 1/2 (therefore, uncountable).

Quantitative Properties

Law of the Iterated Logarithm

$$\limsup_{t\to+\infty} \frac{|w(t)|}{\sqrt{2t\log\log t}} = 1, \quad \text{almost surely.}$$

Modulus of Continuity

Local modulus of continuity:

$$\limsup_{\varepsilon\to0+} \frac{|w(\varepsilon)|}{\sqrt{2\varepsilon\log\log(1/\varepsilon)}} = 1, \qquad \text{almost surely.}$$

Global modulus of continuity (Lévy):

$$\limsup_{\varepsilon\to0+} \sup_{0\le s<t\le1,\, t-s\le\varepsilon} \frac{|w(s)-w(t)|}{\sqrt{2\varepsilon\log(1/\varepsilon)}} = 1, \textit{almost surely.}$$

Local Time

The image of the Lebesgue measure on [0, t] under the map w (the pushforward measure) has a density $L_t(\cdot)$. Thus,

$$\int_0^t f(w(s))\mathrm{d}s = \int_{-\infty}^{+\infty} f(x)L_t(x)\mathrm{d}x$$

for a wide class of functions f (namely: all continuous functions; all locally integrable functions; all non-negative measurable functions). The density L_t is (more exactly, can and will be chosen to be) continuous. The number $L_t(x)$ is called the local time at x of w on [0, t]. It is strictly positive for all x of the interval (a, b) where a and b are the least and the greatest value of w on [0, t], respectively. (For x outside this interval the local time evidently vanishes.) Treated as a function of two variables x and t, the local time is still continuous. Treated as a function of t (while x is fixed), the local time is a singular function corresponding to a nonatomic measure on the set of zeros of w.

These continuity properties are fairly non-trivial. Consider that the local time can also be defined (as the density of the pushforward measure) for a smooth function. Then, however, the density is discontinuous, unless the given function is monotone. In other words, there is a conflict between good behavior of a function and good behavior of its local time. In this sense, the continuity of the local time of the Wiener process is another manifestation of non-smoothness of the trajectory.

Related Processes

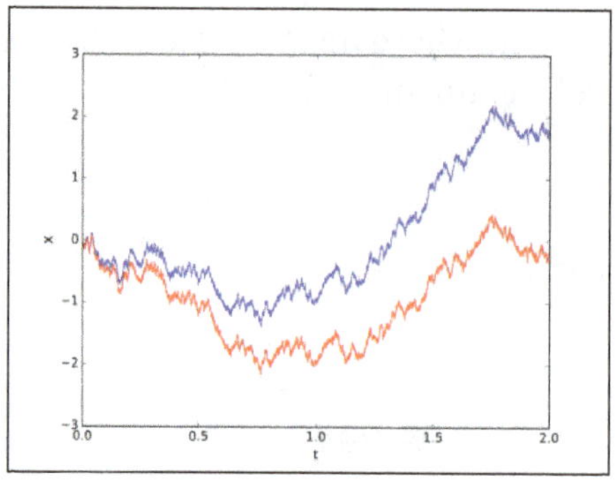

Wiener processes with drift (blue) and without drift (ref).

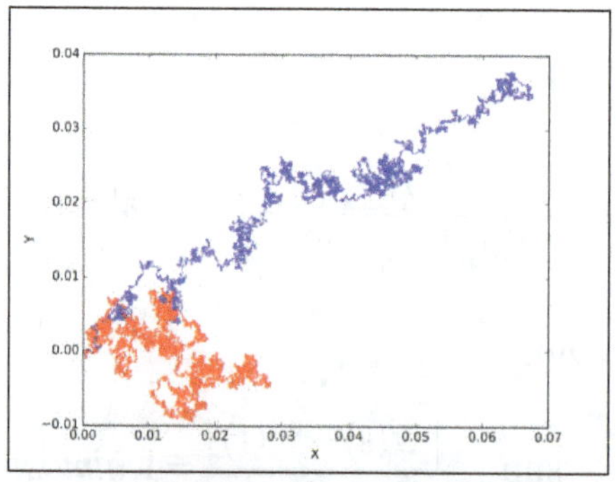

2D Wiener processes with drift (blue) and without drift (ref).

The generator of a Brownian motion is ½ times the Laplace–Beltrami operator. The image above is of the Brownian motion on a special manifold: the surface of a sphere.

The stochastic process defined by:

$$X_t = \mu t + \sigma W_t$$

is called a Wiener process with drift μ and infinitesimal variance σ^2. These processes exhaust continuous Lévy processes.

Two random processes on the time interval [0, 1] appear, roughly speaking, when conditioning the Wiener process to vanish on both ends of [0,1]. With no further conditioning, the process takes both positive and negative values on [0, 1] and is called Brownian bridge. Conditioned also to stay positive on (0, 1), the process is called Brownian excursion. In both cases a rigorous treatment involves a limiting procedure, since the formula $P(A|B) = P(A \cap B)/P(B)$ does not apply when $P(B) = 0$.

A geometric Brownian motion can be written:

$$\mu t - \text{——} + \sigma W$$

It is a stochastic process which is used to model processes that can never take on negative values, such as the value of stocks.

The stochastic process:

$$X_t = e^{-t} W_{e^{2t}}$$

is distributed like the Ornstein–Uhlenbeck process.

The time of hitting a single point $x > 0$ by the Wiener process is a random variable with the Lévy distribution. The family of these random variables (indexed by all positive numbers x) is a left-continuous modification of a Lévy process. The right-continuous modification of this process is given by times of first exit from closed intervals $[0, x]$.

The local time $L = (L^x_t)_{x \in \mathbb{R}, t \geq 0}$ of a Brownian motion describes the time that the process spends at the point x. Formally:

$$L^x(t) = \int_0^t \delta(x - B_t)ds$$

where δ is the Dirac delta function. The behaviour of the local time is characterised by Ray–Knight theorems.

Brownian Martingales

Let A be an event related to the Wiener process (more formally: a set, measurable with respect to the Wiener measure, in the space of functions), and X_t the conditional probability of A given the Wiener process on the time interval $[0, t]$ (more formally: the Wiener measure of the set of trajectories whose concatenation with the given partial trajectory on $[0, t]$ belongs to A). Then the process X_t is a continuous martingale. Its martingale property follows immediately from the definitions, but its continuity is a very special fact – a special case of a general theorem stating that all Brownian martingales are continuous. A Brownian martingale is, by definition, a martingale adapted to the Brownian filtration; and the Brownian filtration is, by definition, the filtration generated by the Wiener process.

Integrated Brownian Motion

The time-integral of the Wiener process

$$W^{(-1)}(t) := \int_0^t W(s)ds:$$

is called integrated Brownian motion or integrated Wiener process. It arises in many applications and can be shown to have the distribution $N(0, t^3/3)$, calculus lead using the fact that the covariance of the Wiener process is $t \wedge s = \min(t,s)$.

Time Change

Every continuous martingale (starting at the origin) is a time changed Wiener process.

Example: $2W_t = V(4t)$ where V is another Wiener process (different from W but distributed like W).

Example. $W_t^2 - t = V_{A(t)}$ where $A(t) = 4\int_0^t W_s^2 \, ds$ and V is another Wiener process.

In general, if M is a continuous martingale then $M_t - M_0 = V_{A(t)}$ where $A(t)$ is the quadratic variation of M on [0, t], and V is a Wiener process.

Corollary. Let M_t be a continuous martingale,

$$M_\infty^- = \liminf_{t\to\infty} M_t,$$

$$M_\infty^+ = \limsup_{t\to\infty} M_t.$$

Then only the following two cases are possible:

$$-\infty < M_\infty^- = M_\infty^+ < +\infty,$$

$$-\infty = M_\infty^- < M_\infty^+ = +\infty;$$

other cases (such as $M_\infty^- = M_\infty^+ = +\infty$, $M_\infty^- < M_\infty^+ < +\infty$ etc.) are of probability 0.

Especially, a nonnegative continuous martingale has a finite limit (as $t \to \infty$) almost surely.

All stated (in this subsection) for martingales holds also for local martingales.

Change of Measure

A wide class of continuous semimartingales (especially, of diffusion processes) is related to the Wiener process via a combination of time change and change of measure.

Using this fact, the qualitative properties stated above for the Wiener process can be generalized to a wide class of continuous semimartingales.

Complex-valued Wiener Process

The complex-valued Wiener process may be defined as a complex-valued random process of the form $Z_t = X_t + iY_t$ where X_t, Y_t are independent Wiener processes (real-valued).

Self-similarity

Brownian scaling, time reversal, time inversion: the same as in the real-valued case.

Rotation invariance: for every complex number c such that $|c| = 1$ the process cZ_t is another complex-valued Wiener process.

Time Change

If f is an entire function then the process $f(Z_t) - f(0)$ is a time-changed complex-valued Wiener process.

Example: $Z_t^2 = (X_t^2 - Y_t^2) + 2X_tY_ti = U_{A(t)}$ where;

$$A(t) = 4\int_0^t |Z_s|^2 \, ds$$

and U is another complex-valued Wiener process.

In contrast to the real-valued case, a complex-valued martingale is generally not a time-changed complex-valued Wiener process. For example, the martingale $2X_t + iY_t$ is not (here X_t, Y_t are independent Wiener processes, as before).

Ornstein–Uhlenbeck Process

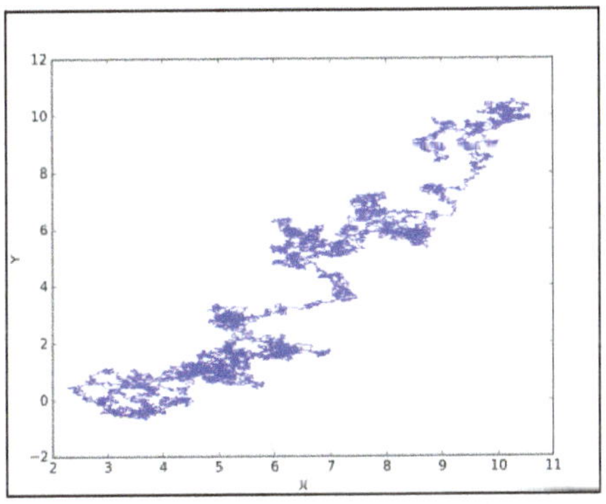

A simulation with θ = 1.0, σ = 300 and μ = (0, 0). Initially at the position (10, 10), the particle tends to move to the central point μ.

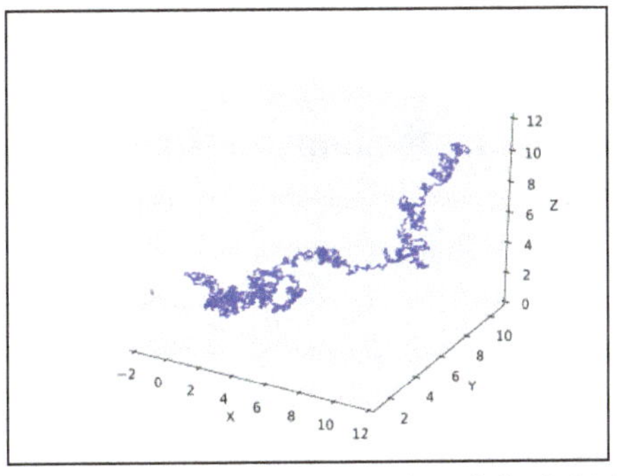

A 3D simulation with θ = 1.0, σ = 300, μ = (0, 0, 0) and the initial position (10, 10, 10).

In mathematics, the Ornstein–Uhlenbeck process (named after Leonard Ornstein and George Eugene Uhlenbeck), is a stochastic process that, roughly speaking, describes the velocity of a massive

Brownian particle under the influence of friction. The process is stationary Gauss–Markov process (which means that it is both a Gaussian and Markovian process), and is the only nontrivial process that satisfies these three conditions, up to allowing linear transformations of the space and time variables. Over time, the process tends to drift towards its long-term mean: such a process is called mean-reverting.

The process can be considered to be a modification of the random walk in continuous time, or Wiener process, in which the properties of the process have been changed so that there is a tendency of the walk to move back towards a central location, with a greater attraction when the process is further away from the centre. The Ornstein–Uhlenbeck process can also be considered as the continuous-time analogue of the discrete-time AR(1) process.

Representation Via a Stochastic Differential Equation

An Ornstein–Uhlenbeck process, x_t, satisfies the following stochastic differential equation:

$$dx_t = \theta(\mu - x_t)dt + \sigma dW_t$$

where $\theta > 0$, μ and $\sigma > 0$ are parameters and W_t denotes the Wiener process.

The above representation can be taken as the primary definition of an Ornstein–Uhlenbeck process or sometimes also mentioned as the Vasicek model.

Fokker–Planck Equation Representation

The probability density function $f(x, t)$ of the Ornstein–Uhlenbeck process satisfies the Fokker–Planck equation,

$$\frac{\partial f}{\partial t} = \theta \frac{\partial}{\partial x}[(x-\mu)f] + \frac{\sigma^2}{2}\frac{\partial^2 f}{\partial x^2}$$

The Green function of this linear parabolic partial differential equation, taking $\mu = 0$ and $D = \sigma^2 / 2$ for simplicity, and the initial condition consisting of a unit point mass at location is,

$$f(x,t) = \sqrt{\frac{\theta}{2\pi D(1-e^{-2\theta t})}} \exp\left\{\frac{-\theta}{2D}\left[\frac{(x-ye^{-\theta t})^2}{1-e^{-2\theta t}}\right]\right\}$$

The stationary solution of this equation is the limit for time tending to infinity which is a Gaussian distribution with mean μ and variance $\sigma^2 / (2\theta)$,

$$f_s(x) = \sqrt{\frac{\theta}{\pi\sigma^2}} e^{-\theta(x-\mu)^2/\sigma^2}.$$

Application in Physical Sciences

The Ornstein–Uhlenbeck process is a prototype of a noisy relaxation process. Consider for example a Hookean spring with spring constant k whose dynamics is highly overdamped with friction coefficient γ. In the presence of thermal fluctuations with temperature T, the length $x(t)$ of the spring will fluctuate stochastically around the spring rest length x_0; its stochastic dynamic is described by an Ornstein–Uhlenbeck process with:

$$\theta = k / \gamma,$$
$$\mu = x_0,$$
$$\sigma = \sqrt{2k_BT / \gamma},$$

where σ is derived from the Stokes–Einstein equation $D = \sigma^2 / 2 = k_BT / \gamma$ for the effective diffusion constant.

In physical sciences, the stochastic differential equation of an Ornstein–Uhlenbeck process is rewritten as a Langevin equation,

$$\dot{x}(t) = -\frac{k}{\gamma}(x(t) - x_0) + \xi(t)$$

where $\xi(t)$ is white Gaussian noise with $\langle \xi(\mathrm{t}_1)\xi(\mathrm{t}_2)\rangle = 2\mathrm{k_B T} / \gamma\delta(\mathrm{t}_1 - \mathrm{t}_2)$.

At equilibrium, the spring stores an average energy $\langle E\rangle = k\langle (x - x_0)^2\rangle / 2 = k_BT / 2$ in accordance with the equipartition theorem.

Application in Financial Mathematics

The Ornstein–Uhlenbeck process is one of several approaches used to model (with modifications) interest rates, currency exchange rates, and commodity prices stochastically. The parameter μ represents the equilibrium or mean value supported by fundamentals; σ the degree of volatility around it caused by shocks, and θ the rate by which these shocks dissipate and the variable reverts towards the mean. One application of the process is a trading strategy known as pairs trade.

Mathematical Properties

The Ornstein–Uhlenbeck process is an example of a Gaussian process that has a bounded variance and admits a stationary probability distribution, in contrast to the Wiener process; the difference between the two is in their "drift" term. For the Wiener process the drift term is constant, whereas for the Ornstein–Uhlenbeck process it is dependent on the current value of the process: if te current value of the process is less than the (long-term) mean, the drift will be positive; if the current value of the process is greater than the (long-term) mean, the drift will be negative. In other words, the mean acts as an equilibrium level for the process. This gives the process its informative name, "mean-reverting." The stationary (long-term) variance is given by,

$$\mathrm{var}(x_t) = \frac{\sigma^2}{2\theta}.$$

The Ornstein–Uhlenbeck process is the continuous-time analogue of the discrete-time AR(1) process.

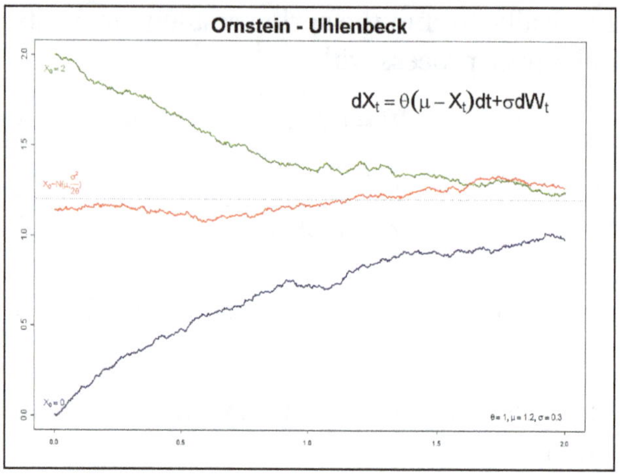

three sample paths of different OU-processes with $\theta = 1$, $\mu = 1.2$, $\sigma = 0.3$:
blue: initial value $a = 0$ (a.s.)
green: initial value $a = 2$ (a.s.)
red: initial value normally distributed so that the process has invariant measure

Solution

This stochastic differential equation is solved by variation of parameters. Changing variable,

$$f(x_t, t) = x_t e^{\theta t}$$

we get

$$\begin{aligned} df(x_t, t) &= \theta x_t e^{\theta t} dt + e^{\theta t} dx_t \\ &= e^{\theta t} \theta \mu dt + \sigma e^{\theta t} dW_t. \end{aligned}$$

Integrating from 0 to t we get

$$x_t e^{\theta t} = x_0 + \int_0^t e^{\theta s} \theta \mu ds + \int_0^t \sigma e^{\theta s} dW_s$$

whereupon we see

$$x_t = x_0 e^{-\theta t} + \mu(1 - e^{-\theta t}) + \sigma \int_0^t e^{-\theta(t-s)} dW_s.$$

Formulas for Moments of Nonstationary Processes

From this representation, the first moment is given by (assuming that x_0 is a constant)

$$E(x_t) = x_0 e^{-\theta t} + \mu(1 - e^{-\theta t})$$

The Itō isometry can be used to calculate the covariance function by:

$$\text{cov}(x_s, x_t) = E[(x_s - E[x_s])(x_t - E[x_t])]$$
$$= E\left[\int_0^s \sigma e^{\theta(u-s)}\, dW_u \int_0^t \sigma e^{\theta(v-t)}\, dW_v\right]$$
$$= \sigma^2 e^{-\theta(s+t)} E\left[\int_0^s e^{\theta u}\, dW_u \int_0^t e^{\theta v}\, dW_v\right]$$
$$= \frac{\sigma^2}{2\theta} e^{-\theta(s+t)} (e^{2\theta \min(s,t)} - 1).$$

Thus if $s < t$ (so that $\min(s, t) = s$), then we have:

$$\text{cov}(x_s, x_t) = \frac{\sigma^2}{2\theta}\left(e^{-\theta(t-s)} - e^{-\theta(t+s)}\right).$$

Alternative Representation for Nonstationary Processes

It is also possible (and often convenient) to represent x_t (unconditionally, i.e. as $t \to \infty$) as a scaled time-transformed Wiener process:

$$x_t = \mu + \frac{\sigma}{\sqrt{2\theta}} e^{-\theta t} W_{e^{2\theta t}}$$

or conditionally (given x_0) as:

$$x_t = x_0 e^{-\theta t} + \mu(1 - e^{-\theta t}) + \frac{\sigma}{\sqrt{2\theta}} e^{-\theta t} W_{e^{2\theta t} - 1}.$$

The time integral of this process can be used to generate noise with a $1/f$ power spectrum.

Scaling Limit Interpretation

The Ornstein–Uhlenbeck process can be interpreted as a scaling limit of a discrete process, in the same way that Brownian motion is a scaling limit of random walks. Consider an urn containing n blue and yellow balls. At each step a ball is chosen at random and replaced by a ball of the opposite colour (equivalently, a ball chosen uniformly at random changes color). Let X_n be the number of blue balls in the urn after n steps. Then $\frac{X_{[nt]} - n/2}{\sqrt{n}}$ converges in law to an Ornstein–Uhlenbeck process as n tends to infinity.

Generalizations

It is possible to extend Ornstein–Uhlenbeck processes to processes where the background driving process is a Lévy process (instead of a simple Brownian motion).These processes are widely studied by Ole Barndorff-Nielsen and Neil Shephard, and others.

In addition, in finance, stochastic processes are used the volatility increases for larger values of X.. In particular, the CKLS (Chan–Karolyi–Longstaff–Sanders) process with the volatility term replaced by $x\ dW$ can be solved in closed form for $\gamma = 1/2$ or 1, as well as for $\gamma = 0$, which corresponds to the conventional OU process.

Random Walk

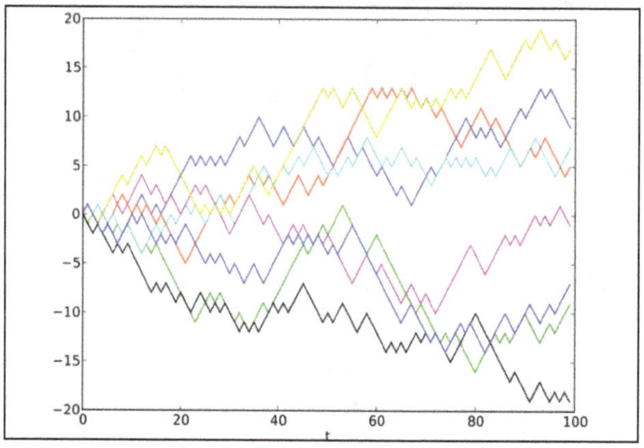

Example of eight random walks in one dimension starting at 0. The plot shows the current position on the line (vertical axis) versus the time steps (horizontal axis).

A random walk is a mathematical object which describes a path that consists of a succession of random steps. For example, the path traced by a molecule as it travels in a liquid or a gas, the search path of a foraging animal, superstring behavior, the price of a fluctuating stock and the financial status of a gambler can all be approximated by random walk models, even though they may not be truly random in reality. As illustrated by those examples, random walks have applications to many scientific fields including ecology, psychology, computer science, physics, chemistry, and biology, and also to economics. Random walks explain the observed behaviors of many processes in these fields, and thus serve as a fundamental model for the recorded stochastic activity. As a more mathematical application, the value of pi can be approximated by the usage of random walk in agent-based modelling environment. The term *random walk* was first introduced by Karl Pearson in 1905.

Various different types of random walks are of interest, which can differ in several ways. The term itself most often refers to a special category of Markov chains or Markov processes, but many time-dependent processes are referred to as random walks, with a modifier indicating their specific properties. Random walks (Markov or not) can also take place on a variety of spaces: commonly studied ones include graphs, others on the integers or the real line, in the plane or in higher-dimensional vector spaces, on curved surfaces or higher-dimensional Riemannian manifolds, and also on groups finite, finitely generated or Lie. The time parameter can also be tinkered with. In the simplest context the walk is in discrete time, that is a sequence of random variables $(Xt) = (X1, X2,...)$ indexed by the natural numbers. However, it is also possible to define random walks which take their steps at random times, and in that case the position X t has to be defined for all times $t \in [0,+\infty]$. Specific cases or limits of random walks include the Lévy flight and diffusion models such as Brownian motion.

Random walks are a fundamental topic in discussions of Markov processes. Their mathematical study has been extensive. Several properties, including dispersal distributions, first-passage or hitting times, encounter rates, recurrence or transience, have been introduced to quantify their behaviour.

Lattice Random Walk

A popular random walk model is that of a random walk on a regular lattice, where at each step the location jumps to another site according to some probability distribution. In a simple random walk, the location can only jump to neighboring sites of the lattice, forming a lattice path. In simple symmetric random walk on a locally finite lattice, the probabilities of the location jumping to each one of its immediate neighbours are the same. The best studied example is of random walk on the d-dimensional integer lattice (sometimes called the hypercubic lattice) . $\mathbb{Z}^d$.

If the state space is limited to finite dimensions, the random walk model is called bordered symmetric random walk and the transition probabilities depend on the location of the state, because on margin and corner states the movement is limited.

One-dimensional Random Walk

An elementary example of a random walk is the random walk on the integer number line, $\mathbb{Z}$, which starts at 0 and at each step moves +1 or −1 with equal probability.

This walk can be illustrated as follows. A marker is placed at zero on the number line and a fair coin is flipped. If it lands on heads, the marker is moved one unit to the right. If it lands on tails, the marker is moved one unit to the left. After five flips, the marker could now be on 1, −1, 3, −3, 5, or −5. With five flips, three heads and two tails, in any order, will land on 1. There are 10 ways of landing on 1 (by flipping three heads and two tails), 10 ways of landing on −1 (by flipping three tails and two heads), 5 ways of landing on 3 (by flipping four heads and one tail), 5 ways of landing on −3 (by flipping four tails and one head), 1 way of landing on 5 (by flipping five heads), and 1 way of landing on −5 (by flipping five tails). See the figure below for an illustration of the possible outcomes of 5 flips.

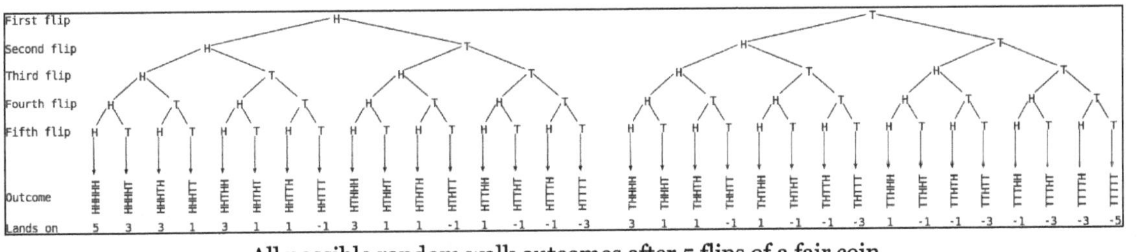

All possible random walk outcomes after 5 flips of a fair coin.

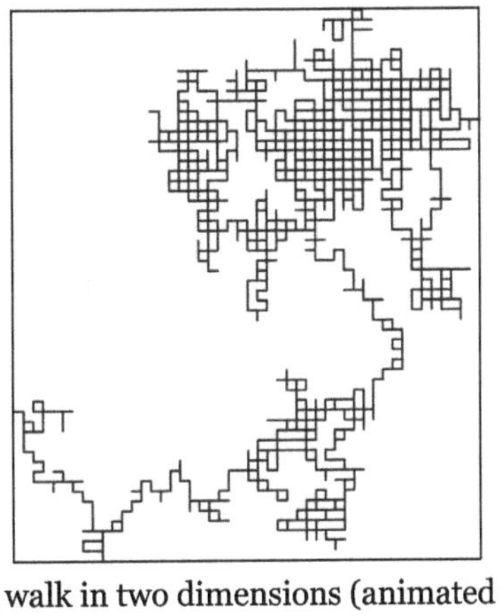

Random walk in two dimensions (animated version).

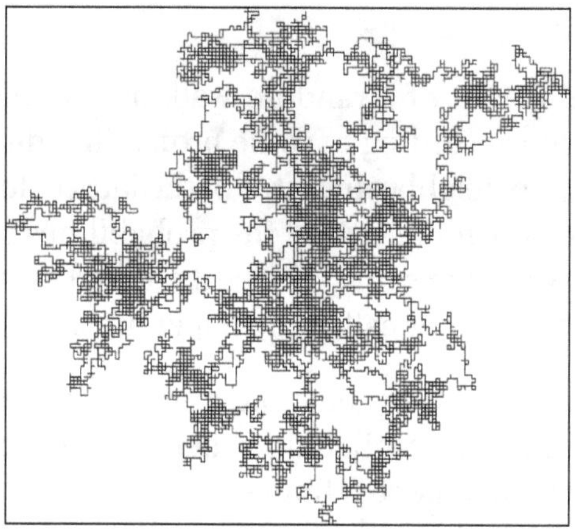

Random walk in two dimensions with 25 thousand steps (animated version).

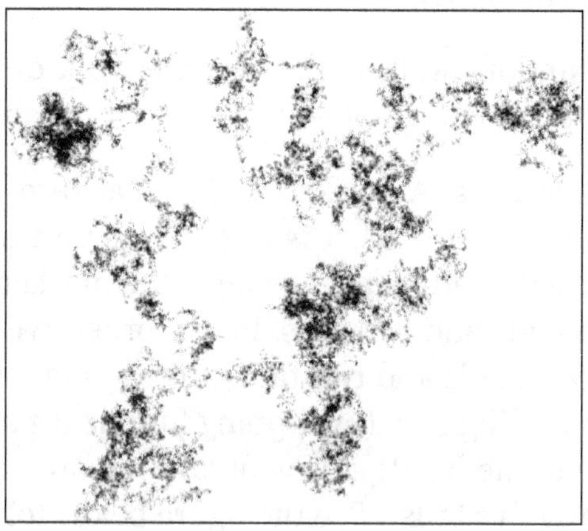

Random walk in two dimensions with two million even smaller steps. This image was generated in such a way that points that are more frequently traversed are darker. In the limit, for very small steps, one obtains Brownian motion.

To define this walk formally, take independent random variables $Z_1, Z_2, \ldots$, where each variable is either 1 or −1, with a 50% probability for either value, and set $S_0 = 0$ and $S_n = \sum Z_j$. The series $\{S_n\}$ is called the simple random walk on $\mathbb{Z}$. This series (the sum of the sequence of −1s and 1s) gives the distance walked, if each part of the walk is of length one. The expectation $E(S_n)$ of S_n is zero. That is, the mean of all coin flips approaches zero as the number of flips increases. This follows by the finite additivity property of expectation:

$$E(S_n) = \sum_{j=1}^{n} E(Z_j) = 0.$$

A similar calculation, using the independence of the random variables and the fact that $E(Z_n^2) = 1$, , shows that:

$$E(S_n^2) = \sum_{i=1}^{n} \sum_{j=1}^{n} E(Z_j Z_i) = n.$$

This hints that $E(|S_n|)$, the expected translation distance after n steps, should be of the order of

$\sqrt{n}$. In fact,

$$\lim_{n\to\infty}\frac{E(|S_n|)}{\sqrt{n}}=\sqrt{\frac{2}{\pi}}.$$

This result shows that diffusion is ineffective for mixing because of the way the square root behaves for large N.

How many times will a random walk cross a boundary line if permitted to continue walking forever? A simple random walk on $\mathbb{Z}$ will cross every point an infinite number of times. This result has many names: the *level-crossing phenomenon*, *recurrence* or the *gambler's ruin*. The reason for the last name is as follows: a gambler with a finite amount of money will eventually lose when playing *a fair game* against a bank with an infinite amount of money. The gambler's money will perform a random walk, and it will reach zero at some point, and the game will be over.

If a and b are positive integers, then the expected number of steps until a one-dimensional simple random walk starting at 0 first hits b or $-a$ is ab. The probability that this walk will hit b before $-a$ is $a/(a+b)$, which can be derived from the fact that simple random walk is a martingale.

Some of the results mentioned above can be derived from properties of Pascal's triangle. The number of different walks of n steps where each step is +1 or −1 is 2^n. For the simple random walk, each of these walks are equally likely. In order for S_n to be equal to a number k it is necessary and sufficient that the number of +1 in the walk exceeds those of −1 by k. The number of walks which satisfy $S_n = k$ is equally the number of ways of choosing $(n - k)/2$ with n is the number of allowed moves, denoted $\binom{n}{(n-k)/2}$. For this to have meaning, it is necessary that n and k be even numbers. Therefore, the probability that $S_n = k$ is equal to $2^{-n}\binom{n}{(n-k)/2}$. By representing entries of Pascal's triangle in terms of factorials and using Stirling's formula, one can obtain good estimates for these probabilities for large values of n.

If the space is confined to $\mathbb{Z}+$ for brevity, the number of ways in which a random walk will land on any given number having five flips can be shown as {0,5,0,4,0,1}.

This relation with Pascal's triangle is demonstrated for small values of n. At zero turns, the only possibility will be to remain at zero. However, at one turn, there is one chance of landing on −1 or one chance of landing on 1. At two turns, a marker at 1 could move to 2 or back to zero. A marker at −1, could move to −2 or back to zero. Therefore, there is one chance of landing on −2, two chances of landing on zero, and one chance of landing on 2.

k	−5	−4	−3	−2	−1	0	1	2	3	4	5
$P[S_0 = k]$						1					
$2P[S_1 = k]$					1		1				
$2^2 P[S_2 = k]$				1		2		1			
$2^3 P[S_3 = k]$			1		3		3		1		

$2^4 P[S_4 = k]$		1		4		6		4		1	
$2^5 P[S_5 = k]$	1		5		10		10		5		1

The central limit theorem and the law of the iterated logarithm describe important aspects of the behavior of simple random walks on $\mathbb{Z}$. In particular, the former entails that as n increases, the probabilities (proportional to the numbers in each row) approach a normal distribution.

As a direct generalization, one can consider random walks on crystal lattices (infinite-fold abelian covering graphs over finite graphs). Actually it is possible to establish the central limit theorem and large deviation theorem in this setting.

As a Markov Chain

A one-dimensional random walk can also be looked at as a Markov chain whose state space is given by the integers $i = 0, \pm 1, \pm 2, \ldots$. For some number p satisfying $0 < p < 1$, , the transition probabilities (the probability $P_{i,j}$ of moving from state i to state j) are given by:

$$P_{i,i+1} = p = 1 - P_{i,i-1}.$$

Higher Dimensions

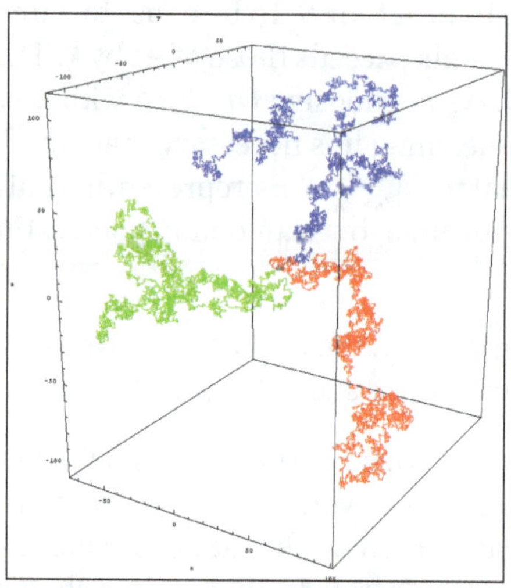

Three random walks in three dimensions.

In higher dimensions, the set of randomly walked points has interesting geometric properties. In fact, one gets a discrete fractal, that is, a set which exhibits stochastic self-similarity on large scales. On small scales, one can observe "jaggedness" resulting from the grid on which the walk is performed. Two books of Lawler referenced below are a good source on this topic. The trajectory of a random walk is the collection of points visited, considered as a set with disregard to *when* the walk arrived at the point. In one dimension, the trajectory is simply all points between the minimum height and the maximum height the walk achieved (both are, on average, on the order of $\sqrt{n}$).

To visualize the two dimensional case, one can imagine a person walking randomly around a city. The city is effectively infinite and arranged in a square grid of sidewalks. At every intersection, the person randomly chooses one of the four possible routes (including the one originally traveled from). Formally, this is a random walk on the set of all points in the plane with integer coordinates.

Will the person ever get back to the original starting point of the walk? This is the 2-dimensional equivalent of the level crossing problem discussed above. It turns out that the person almost surely will in a 2-dimensional random walk, but for 3 dimensions or higher, the probability of returning to the origin decreases as the number of dimensions increases. In 3 dimensions, the probability decreases to roughly 34%.

The asymptotic function for a two dimensional random walk as the number of steps increases is given by a Rayleigh distribution. The probability distribution is a function of the radius from the origin and the step length is constant for each step.

$$P(r) = \frac{2r}{N} e^{-r^2/N}$$

Relation to Wiener Process

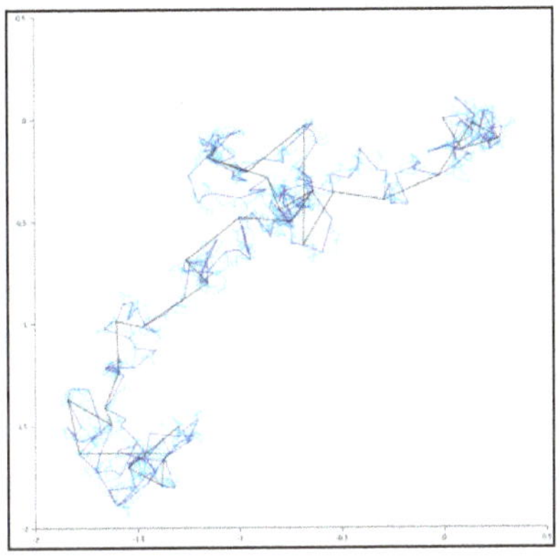

Simulated steps approximating a Wiener process in two dimensions.

A Wiener process is a stochastic process with similar behaviour to Brownian motion, the physical phenomenon of a minute particle diffusing in a fluid. (Sometimes the Wiener process is called "Brownian motion", although this is strictly speaking a confusion of a model with the phenomenon being modeled.)

A Wiener process is the scaling limit of random walk in dimension 1. This means that if you take a random walk with very small steps you get an approximation to a Wiener process (and, less accurately, to Brownian motion). To be more precise, if the step size is ε, one needs to take a walk of length L/ε^2 to approximate a Wiener length of L. As the step size tends to 0 (and the number of steps increases proportionally) random walk converges to a Wiener process in an appropriate sense. Formally, if B is the space of all paths of length L with the maximum topology, and if M is the space of measure over B with the norm topology, then the convergence is in the space M. Similarly, a Wiener process in several dimensions is the scaling limit of random walk in the same number of dimensions.

A random walk is a discrete fractal (a function with integer dimensions; 1, 2, ...), but a Wiener process trajectory is a true fractal, and there is a connection between the two. For example, take a random walk until it hits a circle of radius r times the step length. The average number of steps it performs is r^2. This fact is the *discrete version* of the fact that a Wiener process walk is a fractal of Hausdorff dimension 2.

In two dimensions, the average number of points the same random walk has on the *boundary* of its trajectory is $r^{4/3}$. This corresponds to the fact that the boundary of the trajectory of a Wiener process is a fractal of dimension 4/3, a fact predicted by Mandelbrot using simulations but proved only in 2000 by Lawler, Schramm and Werner.

A Wiener process enjoys many symmetries random walk does not. For example, a Wiener process walk is invariant to rotations, but random walk is not, since the underlying grid is not (random walk is invariant to rotations by 90 degrees, but Wiener processes are invariant to rotations by, for example, 17 degrees too). This means that in many cases, problems on random walk are easier to solve by translating them to a Wiener process, solving the problem there, and then translating back. On the other hand, some problems are easier to solve with random walks due to its discrete nature.

Random walk and Wiener process can be *coupled*, namely manifested on the same probability space in a dependent way that forces them to be quite close. The simplest such coupling is the Skorokhod embedding, but there exist more precise couplings, such as Komlós–Major–Tusnády approximation theorem.

The convergence of a random walk toward the Wiener process is controlled by the central limit theorem, and by Donsker's theorem. For a particle in a known fixed position at $t = 0$, the central limit theorem tells us that after a large number of independent steps in the random walk, the walker's position is distributed according to a normal distribution of total variance:

$$\sigma^2 = \frac{t}{\delta t}\varepsilon^2,$$

where t is the time elapsed since the start of the random walk, ε is the size of a step of the random walk, and δt is the time elapsed between two successive steps.

This corresponds to the Green function of the diffusion equation that controls the Wiener process, which suggests that, after a large number of steps, the random walk converges toward a Wiener process.

In 3D, the variance corresponding to the Green's function of the diffusion equation is:

$$\sigma^2 = 6Dt$$

By equalizing this quantity with the variance associated to the position of the random walker, one obtains the equivalent diffusion coefficient to be considered for the asymptotic Wiener process toward which the random walk converges after a large number of steps:

$$D = \frac{\varepsilon^2}{6\delta t} \text{ (valid only in 3D)}$$

Remark: the two expressions of the variance above correspond to the distribution associated to the vector $\vec{R}$ that links the two ends of the random walk, in 3D. The variance associated to each component R_x, R_y or R_z is only one third of this value (still in 3D).

For 2D:

$$D = \frac{\varepsilon^2}{4\delta t}$$

For 1D:

$$D = \frac{\varepsilon^2}{2\delta t}$$

Gaussian Random Walk

A random walk having a step size that varies according to a normal distribution is used as a model for real-world time series data such as financial markets. The Black–Scholes formula for modeling option prices, for example, uses a Gaussian random walk as an underlying assumption.

Here, the step size is the inverse cumulative normal distribution $\Phi^{-1}(z, \mu, \sigma)$ where $0 \le z \le 1$ is a uniformly distributed random number, and μ and σ are the mean and standard deviations of the normal distribution, respectively.

If μ is nonzero, the random walk will vary about a linear trend. If v_s is the starting value of the random walk, the expected value after n steps will be $v_s + n\mu$.

For the special case where μ is equal to zero, after n steps, the translation distance's probability distribution is given by $N(0, n\sigma^2)$, where $N()$ is the notation for the normal distribution, n is the number of steps, and σ is from the inverse cumulative normal distribution as given above.

Proof: The Gaussian random walk can be thought of as the sum of a series of independent and identically distributed random variables, X_i from the inverse cumulative normal distribution with mean equal zero and σ of the original inverse cumulative normal distribution:

$$Z = \sum_{i=0}^{n} X_i,$$

but we have the distribution for the sum of two independent normally distributed random variables, $Z = X + Y$, is given by:

$$N(\mu_X + \mu_Y, \sigma^2_X + \sigma^2_Y)$$

In our case, $\mu_X = \mu_Y = 0$ and $\sigma^2_X = \sigma^2_Y = \sigma^2$ yield

$$N(0, 2\sigma^2)$$

By induction, for n steps we have

$$Z \sim N(0, n\sigma^2).$$

For steps distributed according to any distribution with zero mean and a finite variance (not necessarily just a normal distribution), the root mean square translation distance after n steps is,

$$\sqrt{E|S_n^2|} = \sigma\sqrt{n}.$$

But for the Gaussian random walk, this is just the standard deviation of the translation distance's distribution after n steps. Hence, if μ is equal to zero, and since the root mean square(rms) translation distance is one standard deviation, there is 68.27% probability that the rms translation distance after n steps will fall between $\pm\,\sigma\sqrt{n}$. Likewise, there is 50% probability that the translation distance after n steps will fall between $\pm\,0.67450\sigma\sqrt{n}$.

Anomalous Diffusion

In disordered systems such as porous media and fractals σ^2 may not be proportional to t but to . The exponent d_w is called the anomalous diffusion exponent and can be larger or smaller than 2. Anomalous diffusion may also be expressed as $\sigma_r^2 \sim Dt^\alpha$ where α is the anomaly parameter. Some diffusions in random environment are even proportional to a power of the logarithm of the time.

Number of Distinct Sites

The number of distinct sites visited by a single random walker $S(t)$ has been studied extensively for square and cubic lattices and for fractals. This quantity is useful for the analysis of problems of trapping and kinetic reactions. It is also related to the vibrational density of states, diffusion reactions processes and spread of populations in ecology. The generalization of this problem to the number of distinct sites visited by N random walkers, $S_N(t)$, has recently been studied for d-dimensional Euclidean lattices. The number of distinct sites visited by N walkers is not simply related to the number of distinct sites visited by each walker.

Applications

Antony Gormley's *Quantum Cloud* sculpture in London was designed by a computer using a random walk algorithm.

As mentioned the range of natural phenomena which have been subject to attempts at description by some flavour random walks is considerable, in particular in physics and chemistry, materials science, biology and various other fields. The following are some specific applications of random walk:

- In financial economics, the "random walk hypothesis" is used to model shares prices and other factors. Empirical studies found some deviations from this theoretical model, especially in short term and long term correlations.
- In population genetics, random walk describes the statistical properties of genetic drift.
- In physics, random walks are used as simplified models of physical Brownian motion and diffusion such as the random movement of molecules in liquids and gases. For example diffusion-limited aggregation. Also in physics, random walks and some of the self interacting walks play a role in quantum field theory.
- In mathematical ecology, random walks are used to describe individual animal movements, to empirically support processes of biodiffusion, and occasionally to model population dynamics.
- In polymer physics, random walk describes an ideal chain. It is the simplest model to study polymers.
- In other fields of mathematics, random walk is used to calculate solutions to Laplace's equation, to estimate the harmonic measure, and for various constructions in analysis and combinatorics.
- In computer science, random walks are used to estimate the size of the Web. In the World Wide Web conference-2006, Bar-Yossef et al. published their findings and algorithms for the same.
- In image segmentation, random walks are used to determine the labels (i.e., "object" or "background") to associate with each pixel. This algorithm is typically referred to as the random walker segmentation algorithm.

In all these cases, random walk is often substituted for Brownian motion.

- In brain research, random walks and reinforced random walks are used to model cascades of neuron firing in the brain.
- In vision science, ocular drift tends to behave like a random walk. According to some authors, fixational eye movements in general are also well described by a random walk.
- In psychology, random walks explain accurately the relation between the time needed to make a decision and the probability that a certain decision will be made.
- Random walks can be used to sample from a state space which is unknown or very large, for example to pick a random page off the internet or, for research of working conditions, a random worker in a given country.
    - When this last approach is used in computer science it is known as Markov Chain

Monte Carlo or MCMC for short. Often, sampling from some complicated state space also allows one to get a probabilistic estimate of the space's size. The estimate of the permanent of a large matrix of zeros and ones was the first major problem tackled using this approach.

- Random walks have also been used to sample massive online graphs such as online social networks.
- In wireless networking, a random walk is used to model node movement.
- Motile bacteria engage in a biased random walk.
- Random walks are used to model gambling.
- In physics, random walks underlie the method of Fermi estimation.
- On the web, the Twitter website uses random walks to make suggestions of who to follow.
- Dave Bayer and Persi Diaconis have proven that 7 riffle shuffles are enough to mix a pack of cards. This result translates to a statement about random walk on the symmetric group which is what they prove, with a crucial use of the group structure via Fourier analysis.

Variants of Random Walks

A number of types of stochastic processes have been considered that are similar to the pure random walks but where the simple structure is allowed to be more generalized. The *pure* structure can be characterized by the steps being defined by independent and identically distributed random variables.

Random Walk on Graphs

A random walk of length *k* on a possibly infinite graph *G* with a root *o* is a stochastic process with random variables $X_1, X_2, \dots, X_k$ such that $X_1 = 0$ and X_{i+1} is a vertex chosen uniformly at random from the neighbors of X_i. Then the number $p_{v,w,k}(G)$ is the probability that a random walk of length *k* starting at *v* ends at *w*. In particular, if *G* is a graph with root *o*, $p_{0,0,2k}$ is the probability that a $2k-$step random walk returns to *o*.

Building on the analogy from the earlier section on higher dimensions, assume now that our city is no longer a perfect square grid. When our person reaches a certain junction he picks between the various available roads with equal probability. Thus, if the junction has seven exits the person will go to each one with probability one seventh. This is a random walk on a graph. Will our person reach his home? It turns out that under rather mild conditions, the answer is still yes. For example, if the lengths of all the blocks are between *a* and *b* (where *a* and *b* are any two finite positive numbers), then the person will, almost surely, reach his home. Notice that we do not assume that the graph is planar, i.e. the city may contain tunnels and bridges. One way to prove this result is using the connection to electrical networks. Take a map of the city and place a one ohm resistor on every block. Now measure the "resistance between a point and infinity". In other words, choose some number *R* and take all the points in the electrical network with distance bigger than *R* from our point and wire them together. This is now a finite electrical network and we may measure the resistance from our point to the wired points. Take R to infinity. The limit is called the *resistance between a point and infinity*. It turns out that the following is true (an elementary proof can be found in the book by Doyle and Snell):

Theorem: *a graph is transient if and only if the resistance between a point and infinity is finite. It is not important which point is chosen if the graph is connected.*

In other words, in a transient system, one only needs to overcome a finite resistance to get to infinity from any point. In a recurrent system, the resistance from any point to infinity is infinite.

This characterization of recurrence and transience is very useful, and specifically it allows us to analyze the case of a city drawn in the plane with the distances bounded.

A random walk on a graph is a very special case of a Markov chain. Unlike a general Markov chain, random walk on a graph enjoys a property called *time symmetry* or *reversibility*. Roughly speaking, this property, also called the principle of detailed balance, means that the probabilities to traverse a given path in one direction or in the other have a very simple connection between them (if the graph is regular, they are just equal). This property has important consequences.

Starting in the 1980s, much research has gone into connecting properties of the graph to random walks. In addition to the electrical network connection described above, there are important connections to isoperimetric inequalities, functional inequalities such as Sobolev and Poincaré inequalities and properties of solutions of Laplace's equation. A significant portion of this research was focused on Cayley graphs of finitely generated groups. In many cases these discrete results carry over to, or are derived from manifolds and Lie groups.

In the context of Random graphs, particularly that of the Erdős–Rényi model, analytical results to some properties of random walkers have been obtained. These include the distribution of first and last hitting times of the walker, where the first hitting time is given by the first time the walker steps into a previously visited site of the graph, and the last hitting time corresponds the first time the walker cannot perform an additional move without revisiting a previously visited site.

A good reference for random walk on graphs is the online book by Aldous and Fill. If the transition kernel $p(x,y)$ is itself random (based on an environment ω) then the random walk is called a "random walk in random environment". When the law of the random walk includes the randomness of the law is called the annealed law; on the other hand, if ω is seen as fixed, the law is called a quenched law.

We can think about choosing every possible edge with the same probability as maximizing uncertainty (entropy) locally. We could also do it globally – in maximal entropy random walk (MERW) we want all paths to be equally probable, or in other words: for each two vertexes, each path of given length is equally probable. This random walk has much stronger localization properties.

Self-interacting Random Walks

There are a number of interesting models of random paths in which each step depends on the past in a complicated manner. All are more complex for solving analytically than the usual random walk; still, the behavior of any model of a random walker is obtainable using computers. Examples include:

- The self-avoiding walk.

The self-avoiding walk of length n on Z^d is the random n-step path which starts at the origin, makes transitions only between adjacent sites in Z^d, never revisits a site, and is chosen uniformly among all such paths. In two dimensions, due to self-trapping, a typical self-avoiding walk is very short, while in higher dimension it grows beyond all bounds. This model has often been used in polymer physics (since the 1960s).

- The loop-erased random walk.
- The reinforced random walk.
- The exploration process.
- The multiagent random walk.

Long-range Correlated Walks

Long-range correlated time series are found in many biological, climatological and economic systems.

- Heartbeat records
- Non-coding DNA sequences
- Volatility time series of stocks
- Temperature records around the globe

Poisson Point Process

In probability, statistics and related fields, a Poisson point process or Poisson process (also called a Poisson random measure, Poisson random point field or Poisson point field) is a type of random mathematical object that consists of points randomly located on a mathematical space. The process has convenient mathematical properties, which has led to it being frequently defined in Euclidean space and used as a mathematical model for seemingly random processes in numerous disciplines such as astronomy, biology, ecology, geology, physics, image processing, and telecommunications.

The Poisson point process is often defined on the real line. For example, in queueing theory it is used to model random events, such as the arrival of customers at a store or phone calls at an exchange, distributed in time. In the plane, the point process, also known as a spatial Poisson process, may represent scattered objects such as transmitters in a wireless network, particles colliding into a detector, or trees in a forest. In this setting, the process is often used in mathematical models and in the related fields of spatial point processes, stochastic geometry, spatial statistics and continuum percolation theory. In more abstract spaces, the Poisson point process serves as an object of mathematical study in its own right.

In all settings, the Poisson point process has the property that each point is stochastically independent to all the other points in the process, which is why it is sometimes called a *purely* or *completely* random process. Despite its wide use as a stochastic model of phenomena representable as points, the inherent nature of the process implies that it does not adequately describe phenomena in which there is sufficiently strong interaction between the points. This has sometimes led to the overuse of the point process in mathematical models, and has inspired other point processes, some of which are constructed via the Poisson point process, that seek to capture this interaction.

The process is named after French mathematician Siméon Denis Poisson despite Poisson never having studied the process. Its name derives from the fact that if a collection of random points in some space forms a Poisson process, then the number of points in a region of finite size is a random variable with a Poisson distribution. The process was discovered independently in several different settings.

The process depends on a single parameter, which, depending on the context, may be a constant, a locally integrable function or, in more general settings, a Radon measure. This parameter represents the expected value of the density of the points in the Poisson process, also known as the intensity; a constant parameter represents constant intensity, a locally integrable function represents variable (location-dependent) intensity, and a Radon measure represents singular intensity. In the first case, the resulting process is called a homogeneous or stationary Poisson point process, and in the other two cases the process is called an inhomogeneous or nonhomogeneous Poisson point process. The word *point* is often omitted, but there are other *Poisson processes* of objects, which, instead of points, consist of more complicated mathematical objects such as lines and polygons, and such processes can be based on the Poisson point process.

History

Poisson Distribution

Despite its name, the Poisson point process was neither discovered nor studied by the French mathematician Siméon Denis Poisson; the name is cited as an example of Stigler's law. The name stems from its inherent relation to the Poisson distribution, derived by Poisson as a limiting case of the binomial distribution. This describes the probability of the sum of n Bernoulli trials with probability p, often likened to the number of heads (or tails) after n biased flips of a coin with the probability of a head (or tail) occurring being p. For some positive constant $\Lambda > 0$, as n increases towards infinity and p decreases towards zero such that the product $np = \Lambda$ is fixed, the Poisson distribution more closely approximates that of the binomial.

Poisson derived the Poisson distribution, published in 1841, by examining the binomial distribution in the limit of p (to zero) and n (to infinity). It only appears once in all of Poisson's work, and the result was not well-known during his time, even though over the following years a number of people would use the distribution without citing Poisson including Philipp Ludwig von Seidel and

Ernst Abbe. The distribution would be studied years after Poisson at the end of the 19th century in a different setting by Ladislaus Bortkiewicz who did cite Poisson and used the distribution with real data to study the number of deaths from horse kicks in the Prussian army.

Discovery

There are a number of claims for early uses or discoveries of the Poisson point process. It has been proposed that the earliest use of the Poisson point process was by John Michell in 1767, a decade before Poisson was born. Michell was interested in the probability a star being within a certain region of another star under the assumption that the stars were "scattered by mere chance", and studied an example consisting of the six brightest stars in the Pleiades, without deriving the Poisson distribution. This work inspired Simon Newcomb to study the problem and to calculate the Poisson distribution as an approximation for the binomial distribution.

At the beginning of the 20th century the Poisson point process would arise independently during the same period in three different situations. In 1909 the Danish mathematician and engineer A.K. Erlang derived the Poisson distribution when developing a mathematical model for the number of incoming phone calls in a finite time interval. Erlang, not at the time aware of Poisson's earlier work, assumed that the number phone calls arriving in each interval of time were independent to each other, and then found the limiting case, which is effectively recasting the Poisson distribution as a limit of the binomial distribution. In 1910 physicists Ernest Rutherford and Hans Geiger, after conducting an experiment in counting the number of alpha particles, published their results in which English mathematician Harry Bateman derived the Poisson probabilities as a solution to a family of differential equations, although Bateman acknowledged that the solutions had been previously solved by others. This experimental work by Rutherford and Geiger partly inspired physicist Norman Campbell who in 1909 and 1910 published two key papers on thermionic noise, also known as shot noise, in vacuum tubes, where it is believed he independently discovered and used the Poisson process. In Campbell's work, he also outlined a form of Campbell's theorem, a key result in the theory of point processes, but Campbell credited the proof to the mathematician G. H. Hardy. The three above discoveries and applications of the Poisson point process has motivated some to say that 1909 should be considered the discovery year of the Poisson point process.

Early Applications

The years after 1909 led to a number of studies and applications of the Poisson point process, however, its early history is complex, which has been explained by the various applications of the process in numerous fields by biologists, ecologists, engineers and others working in the physical sciences. The early results were published in different languages and in different settings, with no standard terminology and notation used. For example, in 1922 Swedish chemist and Nobel Laureate Theodor Svedberg proposed a model in which a spatial Poisson point process is the underlying process in order to study how plants are distributed in plant communities. A number of mathematicians started studying the process in the early 1930s, and important contributions were made by Andrey Kolmogorov, William Feller and Aleksandr Khinchin, among others. As an application, Kolmogorov used a spatial Poisson point process to model the formation of crystals in metals. In the field of teletraffic engineering, where a lot of the early researchers were Danes, such as Erlang, and Swedes, mathematicians and statisticians studied and used Poisson and other point processes.

History of Terms

The Swede Conny Palm in his 1943 dissertation studied the Poisson and other point processes in the one-dimensional setting by examining them in terms of the statistical or stochastic dependence between the points in time. In his work exists the first known recorded use of the term *point process* as *Punktprozesse* in German.

It is believed that William Feller, who also made the term random variable popular over a competing term *chance variable* through a coin flip with Joseph Doob, was the first in print to refer to it as the *Poisson process* in a 1940 paper. Although the Swedish statistician Ove Lundberg used the term *Poisson process* in his 1940 PhD dissertation, in which Feller was acknowledged as an influence, it has been claimed that Feller coined the term before 1940. It has been remarked that both Feller and Lundberg used the term as though it were well-known, implying it was already in spoken use. Feller worked from 1936 to 1939 alongside Swedish mathematician and statistician Harald Cramér at Stockholm University, where Lundberg was a PhD student under Cramér who did not use the term *Poisson process* in a book by him, finished in 1936, but did in subsequent editions, which his has led to the speculation that the term *Poisson process* was coined sometime between 1936 and 1939 at the Stockholm University.

Overview of Definitions

The Poisson point process is one of the most studied point processes, in both the field of probability and in more applied disciplines concerning random phenomena, due to its convenient properties as a mathematical model as well as being mathematically interesting. Depending on the setting, the process has several equivalent definitions as well definitions of varying generality owing to its many applications and characterizations. It may be defined, studied and used in one dimension (on the real line) where it can be interpreted as a counting process or part of a queueing model; in higher dimensions such as the plane where it plays a role in stochastic geometry and spatial statistics; or on more abstract mathematical spaces. Consequently, the notation, terminology and level of mathematical rigour used to define and study the Poisson point process and points processes in general vary according to the context. Despite its different forms and varying generality, the Poisson point process has two key properties.

First Key Property: Poisson Distributed Number of Points

The Poisson point process is related to the Poisson distribution, which implies that the probability of a Poisson random variable N is equal to n is given by:

$$P\{N = n\} = \frac{\Lambda^n}{n!} e^{-\Lambda}$$

where $n!$ denotes n factorial and Λ is the single Poisson parameter that is used to define the Poisson distribution. If a Poisson point process is defined on some underlying mathematical space, called a *state space* or *carrier space*, then the number of points in a bounded region of the space will be a Poisson random variable with some parameter whose form will depend on the setting.

Second key property: Complete Independence

The other key property is that for a collection of disjoint and bounded subregions of the underlying space, the number of points in each bounded subregion will be completely independent of all the others. This property is known under several names such as *complete randomness, complete independence*, or *independent scattering* and is common to all Poisson point processes. In other words, there is a lack of interaction between different regions and the points in general, which motivates the Poisson process being sometimes called a *purely* or *completely* random process.

Different Definitions

The Poisson point process is often defined on the real line in the homogeneous setting, and then extended to a more general settings with more mathematical rigour. For all the instances of the Poisson point process, the two key properties of the Poisson distribution and complete independence play an important role.

Homogeneous Poisson Point Process

If a Poisson point process has a constant parameter, say, λ, then it is called a homogeneous or stationary Poisson point process. The parameter, called *rate* or *intensity*, is related to the expected (or average) number of Poisson points existing in some bounded region. In fact, the parameter λ can be interpreted as the average number of points per some unit of extent such as length, area, volume, or time, depending on the underlying mathematical space, hence it is sometimes called the *mean density*.

Defined on the Real Line

Consider two real numbers a and b, where $a \le b$, and which may represent points in time. Denote by $N(a,b]$ the random number of points of a homogeneous Poisson point process existing with values greater than a but less than or equal to b, or in other words, the number of points of the process in the interval $(a,b]$. If the points form or belong to a homogeneous Poisson process with parameter $\lambda > 0$, then the probability of n points existing in the above interval $(a,b]$ is given by:

$$P\{N(a,b]=n\}=\frac{[\lambda(b-a)]^n}{n!}e^{-\lambda(b-a)},$$

In other words, $N(a,b]$ is a Poisson random variable with mean $\lambda(b-a)$. Furthermore, the number of points in any two disjoint intervals, say, $(a_1,b_1]$ and $(a_2,b_2]$ are independent of each other, and this extends to any finite number of disjoint intervals. In the queueing theory context, one can consider a point existing (in an interval) as an *event*, but this is different to the word event in the probability theory sense. It follows that λ is the expected number of *arrivals* that occur per unit of time, and it is sometimes called the *rate parameter*.

For a more formal definition of a stochastic process, such as a point process, one can use the Kolmogorov theorem, which in this context gives the joint probability of some number of points existing in each disjoint finite interval. More specifically, let $N(a_i,b_i]$ denote the number of points of (a point process) happening in the half-open interval $(a_i, b_i]$, where the real numbers $a_i < b_i \leq a_{i+1}$. Then for some positive integer k, the homogeneous Poisson point process on the real line with parameter $\lambda > 0$ is defined with the finite-dimensional distribution:

$$P\{N(a_i, b_i] = n_i, i = 1, \ldots, k\} = \prod_{i=1}^{k} \frac{[\lambda(b_i - a_i)]^{n_i}}{n_i!} e^{-\lambda(b_i - a_i)},$$

Key Properties

The above definition has two important features pertaining to the Poisson point processes in general:

- the number of points in each finite interval has a Poisson distribution;
- the number of points in disjoint intervals are independent random variables.

Furthermore, it has a third feature related to just the homogeneous Poisson process:

- the distribution of each interval $(a+t, b+t]$ only depends on the interval's length $b-a$..

In other words, for any finite $t > 0$, the random variable $N(a+t, b+t]$ is independent of t, and, hence, the process is Stationary process, which is why it is sometimes called the *stationary Poisson process*.

Law of Large Numbers

The quantity $\lambda(b_i - a_i)$ can be interpreted as the expected or average number of points occurring in the interval $(a_i, b_i]$, namely:

$$E\{N(a_i, b_i]\} = \lambda(b_i - a_i),$$

where E denotes the expectation operator. In other words, the parameter λ of the Poisson process coincides with the *density* of points. Furthermore, the homogeneous Poisson point process adheres to its own form of the (strong) law of large numbers. More specifically, with probability one:

$$\lim_{t \to \infty} \frac{N(t)}{t} = \lambda,$$

where lim denotes the limit of a function.

Memoryless Property

The distance between two consecutive points of a point process on the real line will be an exponential random variable with parameter λ (or equivalently, mean $1/\lambda$). This implies that the points have the memoryless property: the existence of one point existing in a finite interval does not affect the probability (distribution) of other points existing. This property is directly related to the complete independence of the Poisson process, however, it has no natural equivalence when the Poisson process is defined in higher dimensions.

Orderliness and Simplicity

A stochastic process with stationary increments is sometimes said to be *orderly*, *ordinary* or *regular* if:

$$P\{N(t,t+\delta]>1\}=o(\delta),$$

where little-o notation is used. A point process is called a simple point process when the probability of any of its two points coinciding in the same position (on the underlying state space) is zero. For point processes in general on the real line, the (probability distribution) property of orderliness implies that the process is simple or has the (sample path) property of *simplicity*, which is the case for the homogeneous Poisson point process.

Relationship to Other Processes

On the real line, the Poisson point process is a type of continuous-time Markov process known as a birth-death process (with just births and zero deaths) and is called a *pure* or *simple* birth process. More complicated processes with the Markov property, such as Markov arrival processes, have been defined where the Poisson process is a special case.

Counting Process Interpretation

The homogeneous Poisson point process, when considered on the positive half-line, is sometimes defined as a counting process, which can be denoted as $\{N(t),t\geq 0\}$. A counting process represents the total number of occurrences or events that have happened up to and including time t. A counting process is a Poisson counting process with rate $\lambda>0$ if it has the following three properties:

- $N(0)=0$;
- has independent increments;
- the number of events (or points) in any interval of length t is a Poisson random variable with parameter (or mean) λt.

The last property implies

$$E[N(t)]=\lambda t.$$

The Poisson counting process can also be defined by stating that the time differences between events of the counting process are exponential variables with mean $1/\lambda$. The time differences between the events or arrivals are known as interrarrival or interoccurence times. These two definitions of the Poisson counting process agree with the previous definition of the Poisson point process.

Martingale Characterization

On the real line, the homogeneous Poisson point process has a connection to the theory of martingales via the following characterization: a point process is the homogeneous Poisson point process if and only if:

$$N(-\infty,t]-t,$$

is a martingale.

Restricted to the Half-line

If the homogeneous Poisson point process is considered just on the half-line $[0,\infty)$, which is often the case when t represents time, as it does for the previous counting process, then the resulting process is not truly invariant under translation. In that case the process is no longer stationary, according to some definitions of stationarity.

Applications

There have been many applications of the homogeneous Poisson point process on the real line in an attempt to model seemingly random and independent events occurring. It has a fundamental role in queueing theory, which is the probability field of developing suitable stochastic models to represent the random arrival and departure of certain phenomena. For example, customers arriving and being served or phone calls arriving at a phone exchange can be both studied with techniques from queueing theory. In the original paper proposing the online payment system known as Bitcoin featured a mathematical model based on a homogeneous Poisson point process.

Generalizations

The Poisson counting process or, more generally, the homogeneous Poisson point process on the real line is considered one of the simplest stochastic processes for counting random numbers of points. The process can be generalized in a number of ways. One possible generalization is to extend the distribution of interarrival times from the exponential distribution to other distributions, which introduces the stochastic process known as a renewal process. Another generalization is to define it on higher dimensional spaces such as the plane.

Spatial Poisson Point Process

A spatial Poisson process is a Poisson point process defined in the plane $\mathbf{R}^2$. For its definition, consider a bounded, open or closed (or more precisely, Borel measurable) region B of the plane. Denote by $N(B)$ the (random) number of points of N existing in this region $B \subset \mathbf{R}^2$. If the points belong to a homogeneous Poisson process with parameter $\lambda > 0$, then the probability of n points existing in B is given by:

$$P\{N(B)=n\} = \frac{(\lambda\,|\,B\,|)^n}{n!} e^{-\lambda|B|}$$

where $|\,B\,|$ denotes the area of B.

More formally, for some finite integer $k \geq 1$, consider a collection of disjoint, bounded Borel (measurable) sets $B_1,\ldots,B_k$. Let $N(B_i)$ denote the number of points of existing in $N(B_i)$. Then the homogeneous Poisson point process with parameter B_i has the finite-dimensional distribution

$$P\{N(B_i) = n_i, i = 1,\ldots,k\} = \prod_{i=1}^{k} \frac{(\lambda |B_i|)^{n_i}}{n_i!} e^{-\lambda |B_i|}.$$

Applications

According to one statistical study, the positions of cellular or mobile phone base stations in the Australian city Sydney, pictured above, resemble a Poisson point process, while in many other cities around the world they do not and other point processes are required.

The spatial Poisson point process features prominently in spatial statistics, stochastic geometry, and continuum percolation theory. This process is applied in various physical sciences such as a model developed for alpha particles being detected. In recent years, it has been frequently used to model seemingly disordered spatial configurations of certain wireless communication networks. For example, models for cellular or mobile phone networks have been developed where it is assumed the phone network transmitters, known as base stations, are positioned according to a homogeneous Poisson point process.

Defined in Higher Dimensions

The previous homogeneous Poisson point process immediately extends to higher dimensions by replacing the notion of area with (high dimensional) volume. For some bounded region B of Euclidean space $\mathbf{R}^d$, if the points form a homogeneous Poisson process with parameter $\lambda > 0$, then the probability of n points existing in $B \subset \mathbf{R}^d$ is given by:

$$P\{N(B) = n\} = \frac{(\lambda |B|)^n}{n!} e^{-\lambda |B|}$$

where $|B|$ now denotes the n-dimensional volume of B.. Furthermore, for a collection of disjoint, bounded Borel sets $B_1,\ldots,B_k \subset \mathbf{R}^d$, let $N(B_i)$ denote the number of points of N existing in B_i. Then the corresponding homogeneous Poisson point process with parameter $\lambda > 0$ has the finite-dimensional distribution,

$$P\{N(B_i) = n_i, i = 1,\ldots,k\} = \prod_{i=1}^{k} \frac{(\lambda |B_i|)^{n_i}}{n_i!} e^{-\lambda |B_i|}.$$

Homogeneous Poisson point processes do not depend on the position of the underlying state space through its parameter λ, which implies it is both a stationary process (invariant to translation) and an isotropic (invariant to rotation) stochastic process. Similarly to the one-dimensional case, the homogeneous point process is restricted to some bounded subset of $\mathbf{R}^d$, then depending on some definitions of stationarity, the process is no longer stationary.

Points are Uniformly Distributed

If the homogeneous point process is defined on the real line as a mathematical model for occurrences of some phenomenon, then it has the characteristic that the positions of these occurrences or events on the real line (often interpreted as time) will be uniformly distributed. More specifically, if an event occurs (according to this process) in an interval $(a-b]$ where $a \leq b$, then its location will be a uniform random variable defined on that interval. Furthermore, the homogeneous point process is sometimes called the *uniform* Poisson point process. This uniformity property extends to higher dimensions in the Cartesian coordinate, but it does not hold in other coordinate systems (for example, polar or spherical).

Inhomogeneous Poisson Point Process

The inhomogeneous or nonhomogeneous Poisson point process is a Poisson point process with a Poisson parameter set as some location-dependent function in the underlying space on which the Poisson process is defined. For Euclidean space R^d, this is achieved by introducing a locally integrable positive function $\lambda(x)$, where x is a d-dimensional point located in R^d, such that for any bounded region B the (d – dimensional) volume integral of $\lambda(x)$ over region B is finite. In other words, if this integral, denoted by $\Lambda(B)$, is:

$$\Lambda(B) = \int_B \lambda(x)dx < \infty,$$

where dx is a (d-dimensional) volume element,[c] then for any collection of disjoint bounded Borel measurable sets $B_1, \ldots, B_k$, an inhomogeneous Poisson process with (intensity) function $\lambda(x)$ has the finite-dimensional distribution:

$$P\{N(B_i) = n_i, i = 1, \ldots, k\} = \prod_{i=1}^{k} \frac{(\Lambda(B_i))^{n_i}}{n_i!} e^{-\Lambda(B_i)}.$$

Furthermore, $\Lambda(B)$ has the interpretation of being the expected number of points of the Poisson process located in the bounded region B, namely:

$$\Lambda(B) = E[N(B)].$$

Defined on the Real Line

On the real line, the inhomogeneous or non-homogeneous Poisson point process has mean measure given by a one-dimensional integral. For two real numbers a and b, where $a \leq b$, denote by $N(a,b]$ the number points of an inhomogeneous Poisson process with intensity function $\lambda(t)$ with values greater than a but less than or equal to b. The probability of n points existing in the above interval $(a,b]$ is given by:

$$P\{N(a,b]=n\}=\frac{[\Lambda(a,b)]^n}{n!}e^{-\Lambda(a,b)}.$$

where the mean or intensity measure is:

$$\Lambda(a,b)=\int_a^b \lambda(t)dt,$$

which means that the random variable $N(a,b]$ is a Poisson random variable with mean $E\{N(a,b]\}=\Lambda(a,b)$.

A feature of the one-dimension setting considered useful is that an inhomogeneous Poisson point process can be transformed into a homogeneous by a monotone transformation or mapping, which is achieved with the inverse of Λ.

Counting Process Interpretation

The inhomogeneous Poisson point process, when considered on the positive half-line, is also sometimes defined as a counting process. With this interpretation, the process, which is sometimes written as $\{N(t), t\geq 0\}$, represents the total number of occurrences or events that have happened up to and including time t. A counting process is said to be an inhomogeneous Poisson counting process if it has the four properties:

- $N(0)=0$;
- has independent increments;
- $P\{N(t+h)-N(t)=1\}=\lambda(t)h+o(h)$;
- $P\{N(t+h)-N(t)\geq 2\}=o(h)$,

where $o(h)$ is asymptotic or little-o notation for $o(h)/h\to 0$ as $h\to 0$. In the case of point processes with refractoriness (e.g., neural spike trains) a stronger version of property 4 applies: $P(N(t+h)-N(t)\geq 2)=o(h^2)$.

The above properties imply that $N(t+h)-N(t)$ is a Poisson random variable with the parameter (or mean),

$$E[N(t+h)-N(t)]=\int_t^{t+h}\lambda(s)ds,$$

which implies;

$$E[N(h)]=\int_0^h \lambda(s)ds.$$

Spatial Poisson Point Process

An inhomogeneous Poisson process, just like a homogeneous Poisson process, defined in the plane $\mathbf{R}^2$ is called a spatial Poisson point process. Calculating its intensity measure requires performing an area integral of its intensity function over some region. For example, its intensity function (as a function of Cartesian coordinates x and y) may be:

$$\lambda(x,y)=e^{-(x^2+y^2)},$$

hence it has an intensity measure given by the area integral,

$$\Lambda(B)=\int_B e^{-(x^2+y^2)}dxdy,$$

where B is some bounded region in the plane R^2. The previous intensity function can be re-written, via a change of coordinates, in polar coordinates as:

$$\lambda(r,\theta)=e^{-r^2},$$

which reveals that the intensity function in this example is independent of the angular coordinate θ, or, in other words, it is isotropic or rotationally invariant. The intensity measure is then given by the area integral,

$$\Lambda(B')=\int_{B'} e^{-r^2} rdrdd\theta,$$

where B' is some bounded region in the plane R^2.

In Higher Dimensions

In the plane, $\Lambda(B)$ corresponds to an area integral while in $\mathbf{R}^d$ the integral becomes a (d-dimensional) volume integral.

Applications

The real line, as mentioned earlier, is often interpreted as time and in this setting the inhomogeneous process is used in the fields of counting processes and in queueing theory. Examples of phenomena which have been represented by or appear as an inhomogeneous Poisson point process include:

- Goals being scored in a soccer game.
- Defects in a circuit board.

In the plane, the Poisson point process is of fundamental importance in the related disciplines of stochastic geometry and spatial statistics. This point process is not stationary owing to the fact that its distribution is dependent on the location of underlying space or state space. Hence, it can be used to model phenomena with a density that varies over some region. In other words, the phenomena can be represented as points that have a location-dependent density. Uses for this process as a mathematical model are diverse and have appeared across various disciplines including the study of salmon and sea lice in the oceans, forestry, and search problems.

Interpretation of the Intensity Function

The Poisson intensity function $\lambda(x)$ has an interpretation, considered intuitive, with the volume element dx in the infinitesimal sense: $\lambda(x)dx$ is the infinitesimal probability of a point of a Poisson point process existing in a region of space with volume dx located at x.

For example, given a homogeneous Poisson point process on the real line, the probability of finding a single point of the process in a small interval of width δ is approximately $\lambda\delta x$. In fact, such intuition is how the Poisson point process is sometimes introduced and its distribution derived.

Simple Point Process

If a Poisson point process has an intensity measure that is a locally finite and diffuse (or non-atomic), then it is a simple point process. For a simple point process, the probability of a point existing at a single point or location in the underlying (state) space is either zero or one. This implies that, with probability one, no two (or more) points of a Poisson point process coincide in location in the underlying space.

Simulation

Simulating a Poisson point process on a computer is usually done in a bounded region of space, known as a simulation *window*, and requires two steps: appropriately creating a random number of points and then suitably placing the points in a random manner. Both these two steps depend on the specific Poisson point process that is being simulated.

Step 1: Number of Points

The number of points N in the window, denoted here by W, needs to be simulated, which is done by using a (pseudo)-random number generating function capable of simulating Poisson random variables.

Homogeneous Case

For the homogeneous case with the constant λ, the mean of the Poisson random variable N is set to $\lambda|W|$ where $|W|$ is the length, area or (d-dimensional) volume of W.

Inhomogeneous Case

For the inhomogeneous case, $\lambda|W|$ is replaced with the ($d-$ dimensional) volume integral.

$$\Lambda(W) = \int_W \lambda(x)dx$$

Step 2: Positioning of Points

The second stage requires randomly placing the N points in the window W.

Homogeneous Case

For the homogeneous case in one dimension, all points are uniformly and independently placed in the window or interval W. For higher dimensions in a Cartesian coordinate system, each coordinate is uniformly and independently placed in the window W. If the window is not a subspace of Cartesian space (for example, inside a unit sphere or on the surface of a unit sphere), then the points will not be uniformly placed in W, and suitable change of coordinates (from Cartesian) are needed.

Inhomogeneous Case

For the inhomogeneous, a couple of different methods can be used depending on the nature of the intensity function $\lambda(x)$.. If the intensity function is sufficiently simple, then independent and random non-uniform (Cartesian or other) coordinates of the points can be generated. For example, simulating a Poisson point process on a circular window can be done for an isotropic intensity function (in polar coordinates r and θ), implying it is rotationally variant or independent of θ but dependent on, by a change of variable in r if the intensity function is sufficiently simple.

For more complicated intensity functions, one can use an acceptance-rejection method, which consists of using (or 'accepting') only certain random points and not using (or 'rejecting') the other points, based on the ratio.

$$\frac{\lambda(x_i)}{\Lambda(W)} = \frac{\lambda(x_i)}{\int_W \lambda(x)dx.}$$

where x_i is the point under consideration for acceptance or rejection.

General Poisson Point Process

The Poisson point process can be further generalized to what is sometimes known as the general Poisson point process by using a Radon measure Λ, hence this measure is locally finite. The Radon measure can be atomic, that is it can have atoms at points in the underlying state space, while some researchers assume the converse where the Radon measure Λ is diffuse or non-atomic. If the measure is atomic Λ, then the number of points at x is a Poisson random variable with mean $\Lambda(x)$.

Assuming that the underlying space of the Poisson point process is $\mathbf{R}^d$ (the space can be more general), then $\Lambda(\{x\}) = 0$ for any single point x in $\mathbf{R}^d$ and $\Lambda(B)$ is finite for any bounded subset B of $\mathbf{R}^d$. Then a point process N is a general Poisson point process with intensity Ë if it has the two following properties:

- the number of points in a bounded Borel set B is a Poisson random variable with mean $\Lambda(B)$. In other words, denote the total number of points located in B by $N(B)$, then the probability that the random variable $N(B)$ is equal to n is given by:

$$P\{N(B) = n\} = \frac{(\Lambda(B))^n}{n!} e^{-\Lambda(B)}$$

- the number of points in n disjoint Borel sets forms n independent random variables.

The Radon measure Ë maintains its previous interpretation of being the expected number of points of N located in the bounded region B, namely,

$$\Lambda(B) = E[N(B)].$$

Furthermore, if Λ is absolutely continuous such that it has a density (or more precisely, a Radon–Nikodym density or derivative) with respect to the Lebesgue measure, then for all Borel sets B it can be written as:

$$\Lambda(B)=\int_B \lambda(x)dx,$$

where the density $\lambda(x)$ is known, among other terms, as the intensity function.

Terminology

In addition to the word *point* often being omitted, the terminology of the Poisson point process and point process theory varies, which has been criticized. The homogeneous Poisson (point) process is also called a *stationary* Poisson (point) process, sometimes the *uniform* Poisson (point) process, and in the past it was, by William Feller and others, referred to as a *Poisson ensemble* of points. The term *point process* has been criticized and some authors prefer the term *random point field*, hence the terms *Poisson random point field* or *Poisson point field* are also used. A point process is considered, and sometimes called, a random counting measure, hence the Poisson point process is also referred to as a *Poisson random measure*, a term used in the study of Lévy processes, but some choose to use the two terms for slightly different random objects.

The inhomogenous Poisson point process, as well as be being called *nonhomogeneous* or *non-homogeneous*, is sometimes referred to as the *non-stationary*, *heterogeneous* or *spatially dependent* Poisson (point) process.

The measure Λ is sometimes called the *parameter measure* or *intensity measure* or *mean measure*. If Λ has a derivative or density, denoted by $\lambda(x)$, it may be called the *intensity function* of the general Poisson point process or simply the *rate* or *intensity*, since there are no standard terms. For the homogeneous Poisson point process, the intensity is simply a constant $\lambda>0$, which can be referred to as the *mean rate* or *mean density* or *rate parameter*. For $\lambda=1$, the corresponding process is sometimes referred to as the *standard Poisson* (point) process.

The underlying mathematical space on which the point process, Poisson or other, is defined is known as a *state space* or *carrier space*.

Notation

The notation of the Poisson point process depends on its setting and the field it is being applied in. For example, on the real line, the Poisson process, both homogeneous or inhomogeneous, is sometimes interpreted as a counting process, and the notation $\{N(t),t\geq 0\}$ is used to represent the Poisson process.

Another reason for varying notation is due to the theory of point processes, which has a couple of mathematical interpretations. For example, a simple Poisson point process may be considered as a random set, which suggests the notation $x\in N$, implying that N is a random point belonging to or being an element of the Poisson point process N. Another, more general, interpretation is to consider a Poisson or any other point process as a random counting measure, so one can write the number of points of a Poisson point process N being found or located in some (Borel measurable) region B as $N(B)$, which is a random variable. These different interpretations results in notation being used from mathematical fields such as measure theory and set theory.

For general point processes, sometimes a subscript on the point symbol, for example x, is included so one writes (with set notation) $x_i\in N$ instead of $x\in N$, and x can be used for the dummy variable in integral expressions such as Campbell's theorem, instead of denoting random points. Sometimes an uppercase letter denotes the point process, while a lowercase denotes a point from the processs, so, for example, the point x or x_i belongs to or is a point of the point process X, and be written with set notation as $x \in X$ or $x_i \in X$.

Furthermore, the set theory and integral or measure theory notation can be used interchangeably. For example, for a point process N defined on the Euclidean state space $\mathbf{R}^d$ and a (measurable) function f on $\mathbf{R}^d$, the expression,

$$\int_{\mathbf{R}^d} f(x)N(dx) = \sum_{x_i \in N} f(x_i),$$

demonstrates two different ways to write a summation over a point process. More specifically, the integral notation on the left-hand side is interpreting the point process as a random counting measure while the sum on the right-hand side suggests a random set interpretation.

Functionals and Moment Measures

In probability theory, operations are applied to random variables for different purposes. Sometimes these operations are regular expectations that produce the average or variance of a random variable. Others, such as characteristic functions (or Laplace transforms) of a random variable can be used to uniquely identify or characterize random variables and prove results like the central limit theorem. In the theory of point processes there exist analogous mathematical tools which usually exist in the forms of measures and functionals instead of moments and functions respectively. For measures, often their densities (or Radon-Nikodym derivatives), if they exist, are also expressed with respect to the Lebesgue measure.

Laplace Functionals

For a Poisson point process N with intensity measure Λ, the Laplace functional is given by:

$$L_N(f) = e^{-\int_{\mathbf{R}^d}(1-e^{f(x)})\Lambda(dx)},$$

which for the homogeneous case is:

$$L_N(f) = e^{-\lambda\int_{\mathbf{R}^d}(1-e^{f(x)})dx}.$$

One version of Campbell's theorem involves the Laplace functional of the Poisson point process.

Probability Generating Functionals

The probability generating function of non-negative integer-valued random variable leads to the probability generating functional being defined analogously with respect to any non-negative bounded function v on $\mathbf{R}^d$ such that $0 \le v(x) \le 1$. For a point process N the probability generating functional is defined as:

$$G(v) = E\left[\prod_{x \in N} v(x)\right]$$

where the product is performed for all the points in N. If the intensity measure Λ of N is locally finite, then the G is well-defined for any measurable function u on $\mathbf{R}^d$. For a Poisson point process with intensity measure Λ the generating functional is given by:

$$G(v) = e^{-\int_{\mathbf{R}^d}[1-v(x)]\Lambda(dx)},$$

which in the homogeneous case reduces to,

$$G(v) = e^{-\lambda\int_{\mathbf{R}^d}[1-v(x)]dx}.$$

Moment Measure

For a general Poisson point process with intensity measure Λ the first moment measure is its intensity measure:

$$M^1(B) = \Lambda(B),$$

which for a homogeneous Poisson point process with constant intensity λ means:

$$M^1(B) = \lambda |B|,$$

where $|B|$ is the length, area or volume (or more generally, the Lebesgue measure) of B.

For the Poisson case with measure Λ the second moment measure is:

$$M^2(B) = \Lambda(B) + \Lambda(B)^2.$$

which in the homogeneous case reduces to,

$$M^2(B) = \lambda |B| + (\lambda |B|)^2.$$

Factorial Moment Measure

For a general Poisson point process with intensity measure Λ the n-th factorial moment measure is given by the expression:

$$M^{(n)}(B_1 \times, \ldots, \times B_n) = \prod_{i=1}^{n} [\Lambda(B_i)],$$

where Λ is the intensity measure or first moment measure of N, which for some Borel set B is given by:

$$\Lambda(B) = M^1(B) = E[N(B)].$$

For a homogeneous Poisson point process the n-th factorial moment measure is simply:

$$M^{(n)}(B_1 \times, \ldots, \times B_n) = \lambda^n \prod_{i=1}^{n} |B_i|,$$

where $|B_i|$ is the length, area, or volume (or more generally, the Lebesgue measure) of B_i. Furthermore, the n-th factorial moment density is:

$$\mu^{(n)}(x_1,\ldots,x_n) = \lambda^n.$$

Avoidance Function

The avoidance function or void probability v of a point process N is defined in relation to some set B, which is a subset of the underlying space $\mathbf{R}^d$, as the probability of no points of N existing in B. More precisely, for a test set B, the avoidance function is given by:

$$v(B) = P(N(B) = 0).$$

For a general Poisson point process N with intensity measure Λ, its avoidance function is give by:

$$v(B) = e^{-\Lambda(B)}.$$

Rényi's Theorem

It can be shown that simple point processes are completely characterized by their void probabilities. In other words, complete information of a simple point process is captured entirely in its void probabilities. The case for Poisson process is sometimes known as Rényi's theorem, which is named after Alfréd Rényi who discovered the result for the case of a homogeneous point process in one-dimension.

In onc form, the Rényi's theorem says for a diffuse (or non-atomic) Radon measure Ë on $\mathbf{R}^d$ and a set A is a finite union of rectangles (so not Borel[d]) that if N is a countable subset of $\mathbf{R}^d$ such that:

$$P(N(A) = 0) = v(A) = e^{-\Lambda(A)}$$

then N is a Poisson point process with intensity measure.

Point Process Operations

Mathematical operations can be performed on point processes in order to develop suitable mathematical models. One example of an operation is known as thinning which entails deleting or removing the points of some point process according to a rule, hence creating a new process with the remaining points (the deleted points also form a point process). Another example of a point process operation is superimposing (or combining) point processes into one point process.

One of the reasons why the Poisson point process is often used as model is that, under suitable conditions, when performed on a Poisson point process these operations often produce another (usually different) Poisson point process, demonstrating an aspect of mathematical closure. The operations can also be used to create new point processes, which are then also used as mathematical models for the random placement of certain objects.

Thinning

For the Poisson process, the independent $p(x)-$thinning operations results in another Poisson point process. More specifically, a $p(x)$-thinning operation applied to a Poisson point process with intensity measure Ë gives a point process of removed points that is also Poisson point process N_p with intensity measure Λ_p, which for a bounded Borel set B is given by:

$$\Lambda_p(B) = \int_B p(x)\Lambda(dx)$$

Furthermore, after randomly thinning a Poisson point process, the kept or remaining points also form a Poisson point process, which has the intensity measure,

$$\Lambda_p(B) = \int_B (1 - p(x))\Lambda(dx).$$

The two separate Poisson point processes formed respectively from the removed and kept points are stochastically independent of each other. In other words, if a region is known to contain n kept points (from the original Poisson point process), then this will have no influence on the random number of removed points in the same region. This ability to randomly create two independent Poisson point processes from one is sometimes known as *splitting* the Poisson point process.

Superposition

If there is a countable collection of point processes $N_1, N_2 \ldots,$ then their superposition, or, in set theory language, their union,

$$N = \bigcup_{i=1}^{\infty} N_i,$$

also forms a point process. In other words, any points located in any of the point processes $N_1, N_2 \ldots$ will also be located in the superposition of these point processes.

Superposition Theorem

The Superposition theorem of the Poisson point process, which stems directly from the complete independence property, says that the superposition of independent Poisson point processes $N_1, N_2 \ldots$ with mean measures $\Lambda_1, \Lambda_2, \ldots$ will also be a Poisson point process with mean measure,

$$\Lambda = \sum_{i=1}^{\infty} \Lambda_i.$$

In other words, the union of two (or countably more) Poisson processes is another Poisson process. If a point x is sampled from a countable n union of Poisson processes, then the probability that the point x belongs to the j th Poisson process N_j is given by:

$$P(x \in N_j) = \frac{\Lambda_j}{\sum_{i=1}^{n} \Lambda_i}.$$

Homogeneous Case

In the homogeneous case with constant $\lambda_1, \lambda_2 \ldots,$ the two previous expressions reduce to,

$$\lambda = \sum_{i=1}^{\infty} \lambda_i,$$

and

$$P(x \in N_j) = \frac{\lambda_j}{\sum_{i=1}^{n} \lambda_i}.$$

Clustering

The operation clustering is performed when each point x of some point process N is replaced by another (possibly different) point process. If the original process N is a Poisson point process, then the resulting process N_c is called a Poisson cluster point process.

Random Displacement

A mathematical model may require randomly moving points of a point process to other locations on the underlying mathematical space, which gives rise to a point process operation known as displacement or translation. The Poisson point process has been used to model, for example, the movement of plants between generations, owing to the displacement theorem, which loosely says that the random independent displacement of points of a Poisson point process (on the same underlying space) forms another Poisson point process.

Displacement Theorem

One version of the displacement theorem entails first considering a Poisson point process N on $\mathbf{R}^d$ with intensity function $\lambda(x)$. It is then assumed the points of N are randomly displaced somewhere else in $\mathbf{R}^d$ so that each point's displacement is independent and that the displacement of a point formerly at x is a random vector with a probability density $\rho(x,\cdot)$.[e] Then the new point process N_D is also a Poisson point process with intensity function,

$$\lambda_D(y) = \int_{\mathbf{R}^d} \lambda(x)\rho(x,y)dx,$$

which for the homogeneous case with a constant $\lambda > 0$ means:

$$\lambda_D(y) = \lambda.$$

In other words, after each random and independent displacement of points, the original Poisson point process still exists.

The displacement theorem can be extended such that the Poisson points are randomly displaced from one Euclidean space $\mathbf{R}^d$ to another Euclidean space $\mathbf{R}^{d'}$, where $d' \geq 1$ is not necessarily equal to d.

Mapping

Another property that is considered useful is the ability to map a Poisson point process from one underlying space to another space.

Mapping Theorem

If the mapping (or transformation) adheres to some conditions, then the resulting mapped (or transformed) collection of points also form a Poisson point process, and this result is sometimes referred to as the Mapping theorem. The theorem involves some Poisson point process with mean measure on some underlying space. If the locations of the points are mapped (that is, the point process is transformed) according to some function to another underlying space, then the resulting point process is also a Poisson point process but with a different mean measure Λ'.

More specifically, one can consider a (Borel measurable) function f that maps a point process N with intensity measure Λ from one space S, to another space T in such a manner so that the new point process N' has the intensity measure:

$$\Lambda(B)' = \Lambda(f^{-1}(B))$$

with no atoms, where B is a Borel set and f^{-1} denotes the inverse of the function f. If N is a Poisson point process, then the new process N' is also a Poisson point process with the intensity measure Λ'.

Approximations with Poisson Point Processes

The tractability of the Poisson process means that sometimes it is convenient to approximate a non-Poisson point process with a Poisson one. The overall aim is to approximate the both number of points of some point process and the location of each point by a Poisson point process. There a number of methods that can be used to justify, informally or rigorously, approximating the occurrence of random events or phenomena with suitable Poisson point processes. The more rigorous methods involve deriving upper bounds on the probability metrics between the Poisson and non-Poisson point processes, while other methods can be justified by less formal heuristics.

Clumping Heuristic

One method for approximating random events or phenomena with Poisson processes is called the clumping heuristic. The general heuristic or principle involves using the Poisson point process (or Poisson distribution) to approximate events, which are considered rare or unlikely, of some stochastic process. In some cases these rare events are close to independent, hence a Poisson point process can be used. When the events are not independent, but tend to occur in clusters or *clumps*, then if these clumps are suitably defined such that they are approximately independent of each other, then the number of clumps occurring will be close to a Poisson random variable and the locations of the clumps will be close to a Poisson process.

Stein's Method

Stein's method, a rigorous mathematical technique originally developed for approximating random variables such as Gaussian and Poisson variables, has also been developed and applied to point processes. Stein's method can be used to derive upper bounds on probability metrics, which give way to quantify how different two random mathematical objects vary stochastically, of the Poisson and other point processes. Upperbounds on probability metrics such as total variation and Wasserstein distance have been derived.

Researchers have applied Stein's method to Poisson point processes in a number of ways, such as using Palm calculus. Techniques based on Stein's method have been developed to factor into the upper bounds the effects of certain point process operations such as thinning and superposition. Stein's method has also been used to derive upper bounds on metrics of Poisson and other processes such as the Cox point process, which is a Poisson process with a random intensity measure.

Convergence to a Poisson Point Process

In general, when an operation is applied to a general point process the resulting process is usually not a Poisson point process. For example, if a point process, other than a Poisson, has its points randomly and independently displaced, then the process would not necessarily be a Poisson point process. However, under certain mathematical conditions for both the original point process and the random displacement, it has been shown via limit theorems that if the points of a point process are repeatedly displaced in a random and independent manner, then the finite-distribution of the point process will converge (weakly) to that of a Poisson point process.

Similar convergence results have been developed for thinning and superposition operations that show that such repeated operations on point processes can, under certain conditions, result in the process converging to a Poisson point processes, provided a suitable rescaling of the intensity measure (otherwise values of the intensity measure of the resulting point processes would approach zero or infinity). Such convergence work is directly related to the results known as the Palm–Khinchin[f] equations, which has its origins in the work of Conny Palm and Aleksandr Khinchin, and help explains why the Poisson process can often be used as a mathematical model of various random phenomena.

Generalizations of Poisson Point Processes

The Poisson point process can be generalized by, for example, changing its intensity measure or defining on more general mathematical spaces. These generalizations can be studied mathematically as well as used to mathematically model or represent physical phenomena.

Poisson Point Processes on More General Spaces

For mathematical models the Poisson point process is often defined in Euclidean space, but has been generalized to more abstract spaces and plays a fundamental role in the study of random measures, which requires an understanding of certain mathematical fields such as probability theory, measure theory, topology and functional analysis.

In general, the concept of distance is of practical interest for applications while topological structure is needed for Palm distributions, hence point processes are often defined on mathematical spaces equipped with metrics. The necessity of convergence of sequences requires the space to be complete, which has inspired point processes to be studied on specific complete metric spaces.

Furthermore, every realization of a point process in general can be regarded as a counting measure, which has motivated point processes being considered as random measures. Using the techniques of random measures, the Poisson and other point processes has been defined and studied on a locally compact second countable Hausdorff space.

Cox Point Process

A Poisson point process can be generalized by letting its intensity measure Λ to be also random and independent of the underlying Poisson process, which gives rise to the Cox process or doubly stochastic Poisson process, introduced by David Cox in 1955 under the latter name. The intensity measure may be a realization of random variable or a random field. For example, if the logarithm of the intensity measure is a Gaussian random field, then the resulting process is known as a *log Gaussian Cox process*. More generally, the intensity measures is a realization of a non-negative locally finite random measure. Cox point processes exhibit a *clustering* of points, which can be shown mathematically to be larger than those of Poisson point processes. The generality and tractability of Cox processes has resulted in them being used as models in fields such as spatial statistics and wireless networks.

Marked Poisson Point Process

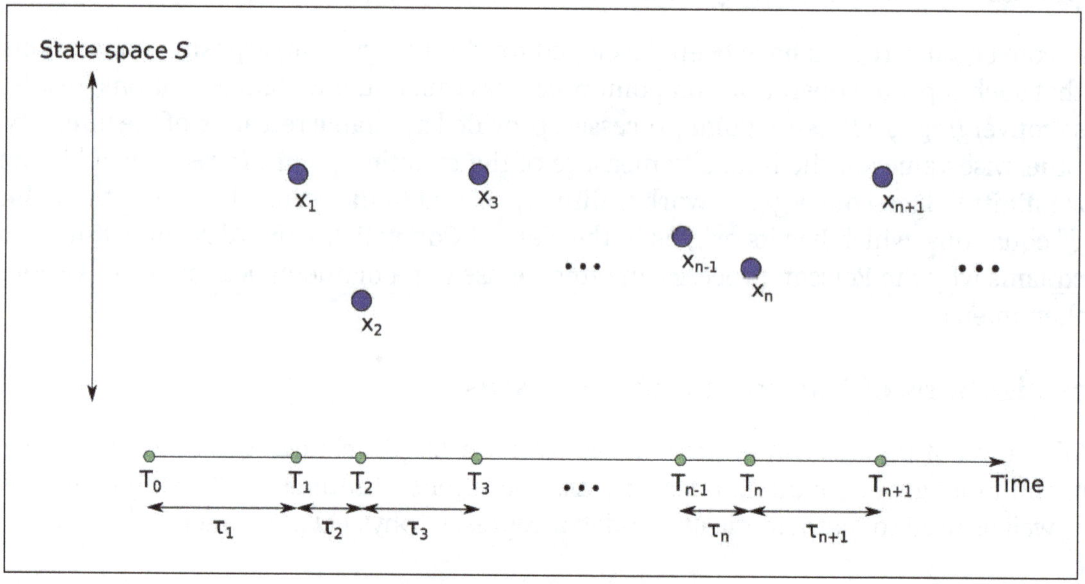

An illustration of a marked point process. If the Poisson process is homogeneous, then the gaps τ_i in the diagram are drawn from an exponential distribution.

For a given point process, each random point of a point process can have a random mathematical object, known as a *mark*, assigned to it. These marks can be as diverse as integers, real numbers, lines, geometrical objects or other point processes. The pair consisting of a point of the point process and its corresponding mark is called a marked point, and all the marked points form a marked point process. It is often assumed that the random marks are independent of each other and identically distributed, which makes the process easier to work with, yet the mark of a point can still depend on the location of its corresponding point in the underlying (state) space. If the underlying point process is a Poisson point process, then one obtains a marked Poisson point process.

Marking Theorem

If a general point process is defined on some mathematical space and the random marks are defined on another mathematical space, then the marked point process is defined on the Cartesian product of these two spaces. For a marked Poisson point process with independent and identically distributed marks, the Marking theorem states that this marked point process is also a (non-marked) Poisson point process defined on the aforementioned Cartesian product of the two mathematical spaces, which is not true for general point processes.

Compound Poisson point process

The compound Poisson point process is formed by adding random values or weights to each point of Poisson point process defined on some underlying state space, so the process is constructed from a marked Poisson point process, where the marks form a collection of independent and identically distributed non-negative random variables. In other words, for each point of the original Poisson process, there is an independent and identically distributed non-negative random variable, and then the compound Poisson process is then formed from the sum of all the random variables corresponding to points of the Poisson process located in a some region of the underlying mathematical space.

IF there is a marked Poisson point processes formed from a Poisson point process N (defined on, for example, $\mathbf{R}^d$) and a collection of independent and identically distributed non-negative marks $\{M_i\}$ such that for each point x_i of the Poisson process N, then there is a non-negative random variable M_i. The resulting compound Poisson process is then:

$$C(B) = \sum_{i=1}^{N(B)} M_i,$$

where $B \subset \mathbf{R}^d$ is a Borel measurable set. If the collection of random variables or marks $\{M_i\}$ are non-negative integer-valued random variables, then the resulting process is called a compound Poisson counting process.

For general random variables, if the compound Poisson point process is formed from a homogeneous Point point process defined on the real line, often representing time, then the resulting compound Poisson process is an example of a Lévy process.

References

- Gardiner, C. Handbook of Stochastic Methods: for Physics, Chemistry and the Natural Sciences, 3rd ed., Springer, 2004, ISBN 3540208828
- Allen, Linda J. S., An Introduction to Stochastic Processes with Applications to Biology, 2nd Edition, Chapman and Hall, 2010, ISBN 1-4398-1882-7
- Durrett, Rick (2010). Probability: Theory and Examples (Fourth ed.). Cambridge: Cambridge University Press. ISBN 978-0-521-76539-8
- Lindsey, J. K. (2004). Statistical Analysis of Stochastic Processes in Time. Cambridge, England: Cambridge University Press. p. 3. ISBN 9780521837415
- Dodge, Yadolah (2006). The Oxford Dictionary of Statistical Terms. Oxford, England: Oxford University Press. p. 335. ISBN 9780199206131

- Kleinert, Hagen (2004). Path Integrals in Quantum Mechanics, Statistics, Polymer Physics, and Financial Markets (4th ed.). Singapore: World Scientific. ISBN 981-238-107-4. (also available online: PDF-files)
- Stark, Henry; Woods, John (2002). Probability and Random Processes with Applications to Signal Processing (3rd ed.). New Jersey: Prentice Hall. ISBN 0-13-020071-9
- Streit, Streit (2010). Poisson Point Processes: Imaging, Tracking, and Sensing. Springer Science& Business Media. ISBN 1441969225
- Daley, Daryl J.; Vere-Jones, David (2003). An Introduction to the Theory of Point Processes: Volume I: Elementary Theory and Methods. Springer. ISBN 1475781091.
- Moller, Jesper; Waagepetersen, Rasmus P. (2003). Statistical Inference and Simulation for Spatial Point Processes. CRC Press. ISBN 1584882654

Permissions

All chapters in this book are published with permission under the Creative Commons Attribution Share Alike License or equivalent. Every chapter published in this book has been scrutinized by our experts. Their significance has been extensively debated. The topics covered herein carry significant information for a comprehensive understanding. They may even be implemented as practical applications or may be referred to as a beginning point for further studies.

We would like to thank the editorial team for lending their expertise to make the book truly unique. They have played a crucial role in the development of this book. Without their invaluable contributions this book wouldn't have been possible. They have made vital efforts to compile up to date information on the varied aspects of this subject to make this book a valuable addition to the collection of many professionals and students.

This book was conceptualized with the vision of imparting up-to-date and integrated information in this field. To ensure the same, a matchless editorial board was set up. Every individual on the board went through rigorous rounds of assessment to prove their worth. After which they invested a large part of their time researching and compiling the most relevant data for our readers.

The editorial board has been involved in producing this book since its inception. They have spent rigorous hours researching and exploring the diverse topics which have resulted in the successful publishing of this book. They have passed on their knowledge of decades through this book. To expedite this challenging task, the publisher supported the team at every step. A small team of assistant editors was also appointed to further simplify the editing procedure and attain best results for the readers.

Apart from the editorial board, the designing team has also invested a significant amount of their time in understanding the subject and creating the most relevant covers. They scrutinized every image to scout for the most suitable representation of the subject and create an appropriate cover for the book.

The publishing team has been an ardent support to the editorial, designing and production team. Their endless efforts to recruit the best for this project, has resulted in the accomplishment of this book. They are a veteran in the field of academics and their pool of knowledge is as vast as their experience in printing. Their expertise and guidance has proved useful at every step. Their uncompromising quality standards have made this book an exceptional effort. Their encouragement from time to time has been an inspiration for everyone.

The publisher and the editorial board hope that this book will prove to be a valuable piece of knowledge for students, practitioners and scholars across the globe.

Index

www.ingramcontent.com/pod-product-compliance
Lightning Source LLC
LaVergne TN
LVHW080718210926
848485LV00003B/484
9781639873364